SUSTAINABILITY

HIGHER EDUCATION'S NEW FUNDAMENTALISM

Hank Campbell, *president, American Council on Science and Health*

Sustainability has become a catch-all term for people and companies trying to attach an ethical halo to their self-identification. The National Academy of Scholars has punctured this mythology and shows how, rather than defending the environment, the sustainability movement primarily results in higher costs—costs that will be borne by those least likely to be able to afford it, but who are duped into thinking they are making a difference.

Willie Soon, *astrophysicist, Cambridge, MA*

Sustainability is a must-read for all truth-seeking and freedom-loving students. If our higher educational institutions are now so willingly corrupted with anti-scientific and pseudo-religious agendas of "cosmic" justices, I can only wish for more determined courage for all to stand up and say "no more." This NAS report will help provide the necessary intellectual ammunition to do so.

Anthony Watts, *meteorologist and publisher of Watts Up With That?*

This book shines a bright spotlight on this new "feel good" ideology, exposing it for what it really is: a magnet for uncritical thinkers who believe the world's problems can be solved with inefficient, overpriced, and unreliable green technology.

David Legates, *Professor of Climatology, University of Delaware*

"The sustainability movement is adversely changing academia through the degradation of intellectual freedom, the corrosion of institutional integrity, and the transition from a "diversity" of thought into a "university" of singular beliefs. Sustainability: Higher Education's New Fundamentalism shines a light on the impact the environmental movement has had on higher education. It is an eye-opener and definitely worth the read."

Matt Briggs, *Statistician*

"The environmental movement's strategic shift from acid rain, deadly pesticides, ozone holes, global cooling, et cetera, and then global warming to sustainability must be acknowledged as brilliant. The former threats are testable and thus make shaky grounds on which to build a movement since the threats could fail to appear. Sustainability, however, means nothing and therefore it can mean anything. No matter what efforts are expended to make a thing "sustainable," more can always be insisted upon. The book which you are about to read makes the sad consequences of this maneuver plain."

Robert Zubrin, *President Pioneer Energy,* **Author of Merchants of Despair**

"Sustainability' is a fundamentally false concept that, denying the central role of human activity in creating new resources, is being used to systematically indoctrinate a generation of young Americans into anti-human modes of thought. By exposing this scandal, Peterson and Wood have performed a major public service."

SUSTAINABILITY
HIGHER EDUCATION'S NEW FUNDAMENTIALISM

First edition published by the National Association of Scholars March 2015.

Second edition with limited updates by Polaris Books September 2016.

Polaris Books
11111 W 8th Ave, Unit A
Lakewood, CO 80215

www.polarisbooks.net

Manufactured in the United States

ABOUT THE NATIONAL ASSOCIATION OF SCHOLARS

Mission

The National Association of Scholars is an independent membership association of academics and others working to sustain the tradition of reasoned scholarship and civil debate in America's colleges and universities. We uphold the standards of a liberal arts education that fosters intellectual freedom, searches for the truth, and promotes virtuous citizenship.

What We Do

We publish a quarterly journal, *Academic Questions*, which examines the intellectual controversies and the institutional challenges of contemporary higher education.

We publish studies of current higher education policy and practice with the aim of drawing attention to weaknesses and stimulating improvements.

Our website presents a daily stream of educated opinion and commentary on higher education and archives our research reports for public access.

NAS engages in public advocacy to pass legislation to advance the cause of higher education reform. We file friend-of-the-court briefs in legal cases, defending freedom of speech and conscience, and the civil rights of educators and students. We give testimony before congressional and legislative committees and engage public support for worthy reforms.

NAS holds national and regional meetings that focus on important issues and public policy debates in higher education today.

Membership

NAS membership is open to all who share a commitment to its core principles of fostering intellectual freedom and academic excellence in American higher education. A large majority of our members are current and former faculty members. We also welcome graduate and undergraduate students, teachers, college administrators, and independent scholars, as well as non-academic citizens who care about the future of higher education.

NAS members receive a subscription to our journal *Academic Questions* and access to a network of people who share a commitment to academic freedom and excellence. We offer opportunities to influence key aspects of contemporary higher education.

Visit our website, www.nas.org, to learn more about NAS and to become a member.

"The protection of planet earth, the survival of all species and sustainability of our ecosystems is more than a mission. It is my religion and my dharma."

- Rajendra Pachauri, former head of the Intergovernmental Panel on Climate Change

TABLE OF CONTENTS

.

TABLES AND KEY ILLUSTRATIONS

ACKNOWLEDGEMENTS

This report would not be possible without the help and support of many individuals and organizations. We are grateful to Mr. Arthur N. Rupe and the Arthur N. Rupe Foundation, whose support enabled us to undertake the research and writing of this report on the campus sustainability movement. We are also grateful to the Weiler Foundation for its support of our economic analysis of the costs of campus sustainability initiatives. Numerous other individuals helped in invaluable ways: Norm Rogers, WillHapper, Lawrence Kogan, Ashley Thorne, and Stanley Kurtz provided information and guidance throughout our research; Liah Greenfeld sponsored Peter's first major paper on sustainability; and Adam Kissel first brought the sustainability movement to our attention.

EXECUTIVE SUMMARY

"Sustainability" is a key idea on college campuses in the United States and the rest of the Western world. To the unsuspecting, sustainability is just a new name for environmentalism. But the word really marks out a new and larger ideological territory in which curtailing economic, political, and intellectual liberty is the price that must be paid now to ensure the welfare of future generations.

This report is the first in-depth critical study of the sustainability movement in higher education. The movement, of course, extends well beyond the college campus. It affects party politics, government bureaucracy, the energy industry, Hollywood, schools, and consumers. But the college campus is where the movement gets its voice of authority, and where it molds the views and commands the attention of young people.

While we take no position in the climate change debate, we focus in this study on how the sustainability movement has distorted higher education. We examine the harm it has done to college curricula and the limits it has imposed on the freedom of students to inquire and to make their own decisions. Our report also offers an anatomy of the campus sustainability movement in the United States. We explain how it came to prominence and how it is organized.

We also examine the financial costs to colleges and universities in their efforts to achieve some of the movement's goals. Often the movement presents its program as saving these institutions money. But we have found that American colleges and universities currently spend more than $3.4 billion per year pursuing their dreams of "sustainability" at a time when college tuitions are soaring and 7.5 percent of recent college graduates are unemployed and another 46 percent underemployed.[1] In addition to the direct costs of the movement, we examine the growing demands by sustainability advocates that colleges and universities divest their holdings in carbon-based energy companies without regard to forgone income or growth in their endowments. What makes "sustainability" so important that institutions facing financial distress are willing to prioritize spending on it? In this report, we examine that question.

Because the idea of "anthropogenic global warming"—or "climate change"—is so closely interwoven with the sustainability movement, we devote a chapter early in the report to laying out the arguments on both

---

1   The basis of the $3.4 billion estimate is given in Chapter 5 of this report. For the unemployment rate see: Heidi Shierholz, Alyssa Davis, and Will Kimball, "The Class of 2014: The Weak Economy Is Idling Too Many Young Graduates," Economic Policy Institute, EPI Briefing Paper #377, May 1, 2014. The Shierholz study uses the term "underemployed" to mean "working part-time" and calculates that 16.8 percent of recent college graduates fit that description. The more common definition of "underemployed" is "working in jobs that generally don't require one to have a college degree." By that definition, 46 percent of recent college graduates are underemployed. See also Catherine Dunn, "Are College Grads Destined For Jobs As Baristas And Clerks? Federal Reserve Economists Explain," *International Business Times*, September 4, 2014. http://www.ibtimes.com/are-college-grads-destined-jobs-baristas-clerks-federal-reserve-economists-explain-1679120

NAS

sides of this debate. The appeal of the sustainability movement depends to a great extent on the belief that the world is experiencing catastrophic warming as a result of human activities that are increasing the amount of carbon dioxide in the atmosphere.

Is this belief warranted? We are neutral on this proposition, but we stand by the principle that all important ideas ought to be open to reasoned debate and careful examination of the evidence. This puts us and others at odds with many in the sustainability movement whose declared position is that the time for debate is over and that those who persist in raising basic questions are "climate deniers." The "debate-is-over" position is itself at odds with intellectual freedom and is why the campus sustainability movement should be examined skeptically.

We support good stewardship of natural resources, but we see in the sustainability movement a hardening of irrational demands to suspend free inquiry in favor of unproven theories of imminent catastrophe. And we see, under the aegis of sustainability, a movement that often takes its bearings from its hostility towards material prosperity, consumerism, free markets, and even democratic self-government.

We offer ten recommendations under three categories:

Respect Intellectual Freedom

1. Create neutral ground. Colleges and universities should be neutral in important and unresolved scientific debates, such as the debate over dangerous anthropogenic global warming. Claims made on the authority of "science" must be made on the basis of transparent evidence and openness to good arguments regardless of their source.

2. Cut the apocalyptic rhetoric. Presenting students with a steady diet of doomsday scenarios undermines liberal education.

3. Maintain civility. Some student sustainability protests have aimed at preventing opponents from speaking.

4. Stop "nudging." Leave students the space to make their own decisions about sustainability, and free faculty members from the implied pressure to embed sustainability into the curricula of unrelated courses.

Uphold Institutional Integrity

5. Withdraw from the ACUPCC. Colleges that have signed the American College and University Presidents' Climate Commitment should withdraw in favor of open-minded debate on the subject.

6. Open the books and pull back the sustainability hires. Make the pursuit of sustainability by colleges financially transparent. The growth of administrative and staff positions in sustainability drives upcosts

NAS

and wrongly institutionalizes advocacy at the expense of education.

7. Uphold environmental stewardship. Campuses need to recover the distinction between real environmental stewardship and a movement that uses the term as a springboard for a much broader agenda.

8. Credential wisely. Curtail the aggrandizement of sustainability as a subject. Sustainability is not a discipline or even a subject area. It is an ideology.

Be Even-Handed

9. Equalize treatment for advocates. Treat sustainability groups on campus under the same rubric as other advocacy groups. They should not enjoy privileged immunity from ordinary rules and special access to institutional resources.

10. Examine motives. College and university boards of trustees should examine demands for divestment from fossil fuels skeptically and with full awareness of the ideological context in which these demands are made.

The sustainability movement has become a major force in American life that has largely escaped serious critical scrutiny. The goal of this report is to change that by examining for the first time the movement's ideological, economic, and practical effects on institutions of higher education.

INTRODUCTION

Sustainability is fast becoming the dominant ideology at colleges and universities in the United States, Britain, and many other parts of the Western world. It is an ideology that harms both the spirit and the substance of liberal education.

Tothe unwary, "sustainability" is the newer name for environmentalism. But the goals of the sustainability movement are different. They go far beyond ensuring clean air and water and protecting vulnerableplants and animals. As an ideology, sustainability takes aim at economic and political liberty. Sustainability pictures economic liberty as a combination of strip mining, industrial waste, and rampant pollution. It pictures political liberty as people voting to enjoy the present, heedless of what it will cost future generations. Sustainability's alternative to economic liberty is a regime of far-reaching regulation that controls virtually every aspect of energy, industry, personal consumption, waste, food, and transportation. Sustainability's alternative to political liberty is control vested in agencies and panels run by experts insulated from elections or other expressions of popular will.

Sustainability's hostility to economic and political liberty is in no sense a secret. Advocates of the movement declare their views at every opportunity. But that part of their message tends not to register with the larger public as much as do the movement's claims about carbon dioxide, global warming, and environmental stewardship. Likewise, the campus sustainability movement (CSM) has not yet come into clear focus for parents, alumni, and the public at large. The purpose of this report is to change that.

If the sustainability movement as a whole has largely escaped critical scrutiny, the campus-based component of the movement has been especially immune to fact-checking and skeptical examination. Since 2006, the year in which the campus sustainability movement became formally organized under the rubric, "The American College and University Presidents' Climate Commitment," there have been no major critical studies of it.

There has, however, been an avalanche of 50,000 books and 200,000 articles expanding on the premises of the movement and advocating for its various goals.[2] Over 100 formal organizations have been created or re-purposed to advance the movement. There are upwards of 50 professional bodies to serve the intellectual and career interests of sustainability experts. There are 1,438 sustainability-focused academic programs at 475 campuses in 65 states and provinces to credential those experts.[3] Hundreds

---

2   World Catalog lists 50,607 books and 229,772 articles on the topic, as of December 1, 2014.

3   According to the Association for the Advancement of Sustainability in Higher Education, as of February 25, 2014. http://www.aashe.org/resources/academic-programs/

NAS

of millions of dollars in private philanthropy have been channeled into sustainability research. Government agencies, too, have poured billions into academic research aligned with the sustainability movement's agenda. The EPA alone has spent more than $333 million in the last 15 years sponsoring sustainability fellowships, predominantly for college and university professors,[4] in addition to another $60 million in sustainability research grants. The National Oceanic and Atmospheric Administration records show more than $3 billion in grants for climate science research since 1998 (more than $89 million in 2014), while the National Institutes of Health has granted in the last four years alone $28 million for research on climate change and another $580 million on "Climate-Related Exposures and Conditions."[5] The National Science Foundation records show more than $1.7 billion since 1998 in sustainability research grants.[6] The National Endowment for the Arts invested $2 million over the same period.[7] The disparity in date ranges available in government grant databases makes direct comparisons difficult. But these numbers indicate an average of $465 million in federal funding for sustainability and climate change research each year—though in recent years government funding for climate research has increased substantially.

In less than a decade, the campus sustainability movement has gone from a minor thread of campus activism to becoming the master narrative of what "liberal education" should seek to accomplish for students and for society as a whole.

In this report, we critique that master narrative.

What Is Sustainability?

The word "sustainability" evokes concerns about conservation, stewardship of the earth's resources, and public policy aimed at ensuring clean air and water for generations to come. Participants in the movement often advise balance, respect for nature's patterns, and protection of natural goods from overuse and from the harms of climate change. They seek to correct market failures by requiring markets to account for externalities such as pollution. And participants say they seek a friendlier economy in which businesses are rewarded on multiple bottom lines—social wellbeing and environmental responsibility in addition to financial rewards.

But the movement is much more than a call for environmental responsibility. It is a summons for fundamental changes in human life—changes that include the imposition of vast new social, political, and economic controls. Sustainability advocates vary among themselves in how far they think these

---

4   "Sustainability Fellowship," EPA, January 2, 2015. http://www.epa.gov/ncer/quickfinder/sustainability_search.html

5   Climate Grants, National Oceanic and Atmospheric Administration, February 20, 2015.

6   Sustainability Grants, National Science Foundation, January 2, 2015. http://www.nsf.gov/awardsearch/simpleSearchResult?queryText=%22sustainability%22&ExpiredAwards=true

7   Sustainability Grants, National Endowment for the Arts, February 20, 2015.

fundamental changes need to go, but a great many of them view "capitalism" as the primary enemy. They see as the root problem the economic and social system that brought modern industrial technology into the world and freed much of humanity from the drudgery of subsistence labor.

For people accustomed to images of clear blue skies and fresh mountain springs when they hear the word "sustainability," the suggestion that sustainability is really a war against the comforts of modern life seems too large a stretch to be true. So let's start with the words of one of the most prominent advocates of the movement, Naomi Klein, whose recent book, *This Changes Everything: Capitalism vs. the Climate*, declares that the environmental crises of today call for no less than the abolition of capitalism. Ms. Klein writes that "the urgency of the climate crisis could form the basis of a powerful mass movement" aimed at protecting "humanity from the ravages of both a savagely unjust economic system and a destabilized climate system."[8]

Ms. Klein blames what she calls "free market fundamentalism" for "overheating the planet." She disparages the idea that markets in general solve most of our problems of scarcity and distribution. And she sees "big business" as inimical to environmental health.

Ms. Klein may sound like an extremist—and indeed she is. In an interview with the Sierra Club, she embraces her role as a "radical" within the movement and acknowledges that her forthright anti-capitalism is outside the comfort zone of many environmentalists: "right now, I'm getting so much pushback for just talking about capitalism."[9] But she is not alone in espousing the view that a truly sustainable society is one that has rid itself of capitalism and, in the process, eliminated the use of fossil fuels. She is one of many in the leadership of this movement who uphold such views. And anyone who seeks to understand the sustainability movement must be prepared to take this first big step: The sustainability movement presents itself as benign concern for the natural environment, but its deeper aim is radical economic transformation.

The more moderate voices in the movement sometimes speak about curbing the excesses of free markets, but its dominant voices view free markets themselves as the cause of environmental disaster. Some even view the existence of private property as the fundamental problem. Let people decide for themselves what to do with the things they own, and people will do foolish things. On campus, for example, they might consume a bottle of water and contribute to our burgeoning landfills. Therefore, on many college campuses, sustainability advocates have succeeded in banning bottled water.

8   Naomi Klein, *This Changes Everything: Capitalism vs. the Climate*, Simon & Schuster, 2014, pg. 8.

9   Steve Hawk, "Capitalism vs. the Planet: Naomi Klein Dares to Discuss the C-word," *Sierra*, January-February 2015. http://www.sierraclub.org/sierra/2015-1-january-february/feature/capitalism-vs-planet

There is a big gap between grand propositions such as "abolish capitalism" and petty actions such as banning bottled water. One of the components of the sustainability movement that we will aim to bring into focus is how it spans global ambitions and micro-administration. Its attempt to do both, however, is an important clue to its character. The environmentalist movement of years gone by focused on getting people to take better care of the natural world. The sustainability movement, by contrast, focuses on convincing people to submit to a regime of nearly total social control.

Even if sustainability has no realistic prospect of attaining the types of totalizing social control its advocates deem desirable, the movement in the meantime fosters a spirit of illiberalism. It does not welcome robust debate on the problems it enunciates. And it has a shameful record of attacking and punishing those it marks out as enemies.

Sustainability thus combines an environmental theme with an economic call to arms and a recipe for harsh and often non-democratic forms of political control. To understand the sustainability movement correctly, however, we need to add one more ingredient: its embrace of identity politics under the rubric of "social sustainability." The basic idea is that a sustainable society must not only rid itself of the penchant for exploiting nature but also of exploitation of oppressed groups of people.

To this end, sustainability reaches into cultural and social institutions, and it demands strict regulation to keep everything in line. Sustainability thus calls for the overthrow of patriarchal systems, misogynist bias, racist prejudice, and traditional marriage norms. It ties social and economic grievances to environmental degradation: women are disproportionately harmed by wars over resource shortages; minorities are more likely to live near landfills and polluting factories; previously colonized and oppressed nations are more likely to be flooded and scorched by global warming; the poor are least responsible for causing climate change but also least able to protect themselves from its effects; traditional marriage (without abortion and strict birth control) overpopulates the globe and mires communities in poverty. Rising global warming and the threat of runaway consumption augment these dangers.

The goal of the sustainability movement is radical transformation of the relation between humanity and nature.

Sustainability, too, warns of wars and rumors of wars over oil, water, food, clean air, rare minerals, precious metals, and land. It castigates free markets for prioritizing individuals over communities—for facilitating the individualism that characterizes free market economics. And sustainability demands strict regulations on corporate behavior and individual

NAS

consumption; education initiatives to train students to conform to sustainable guidelines; government mandates on recycling and carbon caps; and global treaties that consider individuals as citizens primarily of a global, rather than national, commonwealth.

There is in these demands something that borders on totalitarianism. The sustainability movement, of course, doesn't see itself as totalitarian, and it would surely be a mistake to say that it leans towards totalitarian forms of control such as existed in Stalin's Soviet Union or Mao's China, or, in the realm of fiction, Orwell's *1984* or Huxley's *Brave New World*. But that is only to say that sustainability presents a new kind of totalizing impulse. It is a desire for total social control conceived as well-intentioned, kind, and even generous. At least up to a point. Robert F. Kennedy, Jr. recently called for a law to punish people who express skepticism about manmade global warming. He singled out the "Koch brothers" as "contemptible human beings"; called their actions "treasonous"; and declared "they should be in jail." The CEOs of coal companies, he added, "should be in jail for all of eternity."[10]

Intemperate language by an irascible public figure, of course, doesn't necessarily reflect the temper of a large movement which includes many more thoughtful advocates. But to become better acquainted with the sustainability movement is to become familiar with a great deal of such intemperance. Many sustainability advocates, as we will see, do not have much patience for those who

THE NEW WHAT?

"Sustainability" is new. But the new what? The new "venture capital buzzword," says one observer. The "new green," says another. The new American Dream. The new Industrial Revolution. The new space race. The new frontier. In the realm of economics, sustainability is by turns the new currency, the new equity, the new profitability, the new economic bottom line, the new imperative in business, the new path to doing business, the new mantra for success, the new leadership framework, the new "lean," the new quality standard, the new driver of growth and profit. It is the new normal, the new black, the new "reality," the "new vanilla ice cream in the world of work," the new safety. And for the awestruck, it is the new "grand narrative replacing modernism," the new elegance, the new lens, and the new "politics of co-existence in the eco-sphere on which we all depend." We suggest that for many, sustainability is the new fundamentalism, the governing ideology that orders everyday life and gives meaning to public choices.

---

10   Marc Morano, "Update: Video: Robert F. Kennedy Jr. Wants To Jail His Political Opponents – Accuses Koch Brothers of 'Treason' – 'They ought to be serving time for it,'" *Climate Depot*, September 21, 2014. http://www.climatedepot.com/2014/09/21/robert-f-kennedy-jr-wants-to-jail-his-political-opponents-accuses-koch-brothers-of-treason-they-ought-to-be-serving-time-for-it/

NAS

decline to submit to the supposedly superior wisdom of the movement.

So what is sustainability? It is an ideology that attempts to unite environmental activism, anti-capitalism, and a progressive vision of social justice. The three are not equal partners. For some advocates, such as Naomi Klein, the anti-capitalist theme comes first. Environmentalist concerns, for Klein, simply provide a fortunate opportunity to do what she wants to do anyway: get rid of capitalism. For others, such as Bill McKibben, the Middlebury College professor who is at the center of the campus fossil fuel divestment movement, the environment comes first, and economy and society will simply have to adjust to the needs of nature. For still others, such as Mitchell Thomashow, former president of Unity College, social justice is the preeminent aim of sustainability initiatives.

The goal of the sustainability movement is radical transformation of the relation between humanity and nature. To this end, it seeks extreme forms of conservation of natural resources; the virtual elimination of extraction of energy from fossil fuels; a drastic retreat from the forms of mass consumption that are characteristic of the modern world ever since the Industrial Revolution; fundamental redistribution of the world's wealth from richer to poorer countries; the end to industrial development in the underdeveloped parts of the world; and a return wherever possible to subsistence and near-subsistence standards of living. The sustainability movement generally views these goals as impossible to achieve via the forms of governance that prevail among modern nation states. To accomplish their ends, sustainability advocates favor the short-term tactics of international treaties, binding multi-national agreements, and rule-setting by world bodies. In the longer term, they see the need for a form of enlightened despotism in which sustainability-minded rulers would create, impose, and enforce a new set of universal norms.

None of this is to say that the sustainability movement is likely to achieve any of its goals. It is worth taking seriously, however, because as an ideology it is exacting and will continue to exact enormous costs by diverting resources from better ends, co-opting higher education, and instilling in students a profound distaste for political and economic freedom.

Sustainatopians

In explaining what the sustainability movement is and what it hopes to achieve we have, in effect, described a utopian movement. Such movements are endemic to the West, although the name "Utopia" as shorthand for an effort to imagine a human community that has achieved peaceful and benevolent perfection arrived only in 1516, with the publication of Thomas More's book about an imaginary island republic. The utopia of today's sustainability advocates—*sustainatopia*—is also an island of sorts, in the manner suggested by photographs of the Earth from outer space. *This Island Earth*, the title of a 1950s science fiction movie that like many of its kind imagined our world imperiled by an existential threat, captures the balance between the sense of fragility and aspiration.

NAS

Sustainatopians on one hand are always reminding us that the dangers are immense, but they are also evoking a future in which all human conflicts and travails will simply evaporate. How will it happen? The answers are as numerous as the participants in the movement. But it will help to give a few examples.

David Shearman is an Australian-based leading advocate of sustainability and an emeritus professor of medicine at the University of Adelaide, South Australia. Writing in the book *The Climate Change Challenge* and the *Failure of Democracy* with his colleague Joseph Wayne Smith, Shearman comments,

> *Ecological services have little chance of surviving without tight control by law of human activity affecting the environment. This option would be thought of as totalitarian by today's free societies, b but this may be the only solution for us.* [11]

Self-government inevitably falls short, he claims, as men refuse to recognize and prioritize the common good. In fact, democracy proves the worst of all possible forms of government:

> *The institution of liberal democracy fails to adequately address the challenges of the environmental crisis, and by giving an even greater license to greed and individual self-satisfaction, it is potentially a more environmentally destructive social system than most other systems under which humans have lived.* [12]

Plato and Aristotle, along with America's founding fathers, might share Shearman's distaste for a pure democratic regime. Plato preferred a natural, virtuous aristocracy while Aristotle praised a polity for its stability; the American founders aimed at a representative republic meant to "refine and enlarge the public view," as James Madison expressed it in "Federalist No. 10." But the proper regime for a "sustainable" society, Shearman and Smith argue, is a totalitarian dictatorship. A sustainable government is autocratic and clamps down on that dangerous phenomenon, human freedom, and the opportunity for self-government. The model, Shearman suggests in a blog post, is China:

> *The People's Republic of China may hold the key to innovative measures that can both arrest the expected surge in emissions from developing countries and provide developed nations with the means to alternative energy. China curbs individual freedom in favour of communal need. The State will implement those measures seen to be in the common good. ... Crises call for fast and sure action and an educated Chinese leadership could deliver.* [13]

Shearman, to be sure, is a fringe figure. We do not know of other sustainability advocates, at least

11   David Shearman and Joseph Wayne Smith, *The Climate Change Challenge and the Failure of Democracy*, Westport, Connecticut: Pentagon Press, 2008, pg. 71.

12   *Ibid.* pg. 55.

13   David Shearman, "Climate Change and the Failure of the Democratic System," *Doctors for the Environment: Australia*, July 29, 2004.   http://dea.org.au/news/article/climate_change_and_the_failure_of_the_democratic_system

those who have advanced degrees and reside in academia, who go so far towards explicit advocacy of totalitarian government as the solution to climate change. But Shearman isn't necessarily that far from the mainstream.

Consider *New York Times* columnist and best-selling author Thomas L. Friedman. In a series of columns in 2009 and 2010, Friedman argued that the Communist Party in China really does offer an attractive model for addressing global warming. In one column he complained that skeptics in the U.S. had demonized the issue of climate change and had caused the Senate to "scuttle" an energy-climate bill. "While American Republicans were turning climate change into a wedge issue, the Chinese Communists were turning it into a work issue," Friedman wrote. He quoted the chairwoman of the U.S. China Collaboration on Clean Energy who proudly explained, "There is really no debate about climate change in   China."[14]

Friedman also appeared on *Meet the Press* on May 23, 2010, saying that he "fantasized" about making America "China for a day," so that we could "authorize the right solutions" on "everything from the economy to the environment." He then backed away, saying that "I don't want to be China for a second." But, "OK, I want my democracy to work with the same authority, focus and stick-to-itiveness."[15]

The think tank Reason labeled Friedman's view "authoritarian envy."[16] And that is probably what we should take away from both Shearman's and Friedman's expostulations. They and many other global warming alarmists are frustrated that the broader public and the duly elected legislatures in the U.S., U.K., Australia, and other nations have not embraced their cause. They imagine—with rather different degrees of self-awareness—that bypassing the structures of self-governance in favor of coercive authority would provide the "answers" they seek.

The outright totalitarian temptation is not shared by all sustainatopians. Some get there by the back roads. Many sustainability advocates, for example, see themselves as champions of the democratic power that Shearman despises. In the sustainatopia idealized by these activists, the wealthy corporations that buy political influence are regulated down to size, and the people are empowered with true   democracy.

Bill McKibben, whom we have already introduced, was a journalist at the *New Yorker* who launched his environmentalist career with the bestselling *End of Nature* in 1989. Since then he has rocketed to national fame for leading demonstrations against the Keystone XL pipeline and coordinating fossil fuel

---

14   Thomas L. Friedman, "Aren't We Clever?" *New York Times*, September 18, 2010. http://www.nytimes.com/2010/09/19/opinion/19friedman.html?_r=0

15   Matt Welch, "Thomas L. Friedman Wants Us "to be China for a day," to "authorize the right solutions,"" Hit & Run Blog, *Reason*, May 24, 2010. http://reason.com/blog/2010/05/24/thomas-l-friedman-wants-us-to

16   *Ibid.*

divestment campaigns at hundreds of college campuses. He argues against plutocratic rule and in favor of raw "people power." Writing in *Rolling Stone* in August 2012, he noted that,

> *Left to our own devices, citizens might decide to regulate carbon and stop short of the brink; according to a recent poll, nearly two-thirds of Americans would back an international agreement that cut carbon emissions 90 percent by 2050. But we aren't left to our own devices. The Koch brothers, for instance, have a combined wealth of $50 billion, meaning they trail only Bill Gates on the list of richest Americans. They've made most of their money in hydrocarbons, they know any system to regulate carbon would cut those profits, and they reportedly plan to lavish as much as $200 million on this year's elections.*[17]

McKibben's imagining that American citizens, freed from the influence of the evil Koch brothers, "might decide" to embrace his carbon diet is, however, every bit as much a fantasy as Friedman's "China for a day" dream. There is no evidence that Americans would be willing to relinquish their standard of living built on the carbon economy for what would amount to a pre-industrial subsistence-based economy—a dream McKibben has evoked in some of his writing. Or that Americans would gamble on a renewable energy economy based on huge increases in the price of staples and as-yet-uninvented technologies. McKibben's path leads likewise to autocratic impositions, but by a more circuitous route than Shearman's or Friedman's.

McKibben, calling for democratic determination of environmental policies, has built a career rallying people to invite rule by an environmental czar. His grassroots group, 350.org, exhorts people to picket the EPA for stronger carbon dioxide regulations, sit in at the White House to protest the construction of the Keystone XL pipeline, and mass-call Congress to support regulatory enhancement in the name of environmental protection.

For McKibben, environmental degradation stems from the degeneration of democracy into plutocracy. The true voice of the people would demand environmental protection, but first the people need to be liberated from the false consciousness into which they have been lulled by consumerism and the industrial-capitalist complex. This is what most sustainability student activists on college campuses believe as well. They distrust political leaders (and their campus administrative authorities too), fear the free market for trampling individuals to benefit corporations, and cling to rallies and marches as demonstrations of popular power. But what the people want, McKibben and his allies say, is a new policy shift in government, away from capitalism and towards strict regulations.

---

17 Bill McKibben, "Global Warming's Terrifying New Math," *Rolling Stone*, July 19, 2012.
http://www.rollingstone.com/politics/news/global-warmings-terrifying-new-math-20120719

NAS

"Government regulation," of course, covers a wide spectrum, from laws enacted through due legislative process to impositions from agencies acting outside their legitimate mandates. The sustainatopian message is consistently on the side of maximizing the reach of regulatory agencies that deal with environmental issues.

Consider Bill McKibben, writing in *Rolling Stone*: "To make a real difference—to keep us under a temperature increase of two degrees—you'd need to change carbon pricing in Washington, and then use that victory to leverage similar shifts around the world."[18]

Naomi Klein, whom we also introduced above, writes:

> *What we know is that the environmental movement had a series of dazzling victories in the late '60s and in the '70s where the whole legal framework for responding to pollution and to protecting wildlife came into law. It was just victory after victory after victory. And these were what came to be called 'command-and-control' pieces of legislation. It was 'don't do that.' That substance is banned or tightly regulated. It was a top-down regulatory approach. And then it came to a screeching halt when Reagan was elected.*[19]

Al Gore, commenting on the importance of the IPCC, declared that

> *Solving this crisis will require cooperation, and bold action from all sectors—businesses must adopt a more sustainable form of capitalism, governments must regulate emissions and adopt a price on carbon in markets, and people must use their voting power to put a price on climate denial in politics in order to ensure that their nations—and global civilization—move toward a sustainable future.*[20]

Secretary of State John Kerry, in a speech in Indonesia, remarked that

> *Thanks to President Obama's Climate Action Plan, the United States is well on our way to meeting the international commitments to seriously cut our greenhouse gas emissions by 2020, and that's because we're going straight to the largest sources of pollution. We're targeting emissions from transportation— cars, trucks, rail, et cetera—and from power sources, which account together for more than 60 percent of the dangerous greenhouse gases that we release. The President has put in place standards to double the fuel-efficiency of cars on American roads. And we've also proposed curbing carbon pollution from new power plants, and similar regulations are in the works to limit the carbon pollution coming from*

18 *Ibid.*

19   Jason Mark, "Naomi Klein: Big Green Is in Denial," *Salon*, September 5, 2013. http://www.salon.com/2013/09/05/naomi_klein_big_green_groups_are_crippling_the_environmental_movement_partner/

20   Al Gore, "Statement on the Intergovernmental Panel on Climate Change's Working Group III Report," *Al Gore*, April 13, 2014. http://blog.algore.com/2014/04/

power plants that are already up and already running.[21]

Why College?

This report examines the sustainability movement on campus. Clearly the movement itself is much broader. It serves as the basis of political careers; it has provided the plots of so many movies that a new genre has emerged called "cli-fi" that trades in imaginary global climate catastrophes. It is built into the marketing of tens of thousands of consumer products, from automobiles to facial tissues. And it is the conceptual backbone of whole new industries that aim to produce green energy. Why then focus on the college campus?

Because colleges and universities have become the linchpins in this movement. That's where the activists focus their efforts to recruit new adherents; that's where the movement develops its new tactics and ideas; that's where federal research money for sustainability is concentrated; and that's where the movement looks for its intellectual and cultural authority.

We are not alone in thinking this. In its early years, in the 1980s and early 1990s, the sustainability movement was mainly an affair of international development experts and government bureaucrats. Its popular support was meager and it had no significant presence on the American college campus. In 1992, Teresa Heinz and John Kerry, after attending the UN Rio Earth Summit, brainstormed together and decided that the best way to bring the movement to life in the United States would be to build its campus presence. To that end they founded Second Nature, an organization specifically devoted to making sustainability a campus issue. We will tell this story in detail later. But, to a large extent, the sustainability movement we have in the United States today is a product of this campus activism far more than it is the result of international summits, political campaigns, government regulation, or green investments. All of those play significant roles, of course, but the cultural basis of the sustainability movement in the United States is higher education.

Those activists may dream of a time when they will be able to impose their will on all of society. But in the shorter term, they focus their efforts on dominating higher education, both intellectually by precluding the expression of dissent, and socially by enforcing their own standards of behavior.

When top-down regulation falls short, education and training programs encourage people voluntarily to police themselves and their neighbors. Adam Corner, writing for the *Guardian* in the UK, comments that
Over the past two decades, a huge amount of time and effort has been expended trying to

21   John Kerry, "Remarks on Climate Change," Jakarta, Indonesia, February 16, 2014. http://www.state.gov/secretary/remarks/2014/02/221704.htm

understand how to nudge, persuade, cajole or regulate people into more sustainable patterns of behaviour. But in our eagerness to understand the drivers of behaviour, and our enthusiasm for measureable behavioural outcomes, we may have overlooked a critical point: that sustained and substantive behavioural transformations come not from gradually 'reprogramming' our behaviour but from internalising the reasons for doing so.[22]

On campus, students sign sustainability pledges, learn about sustainability during orientation, absorb sustainability in their courses, take mandatory sustainability training in residence life programs, are barraged by paid student peers ("eco-reps") who ask them to alter their behavior, and come to see sustainability—strict environmentalism, social experimentation, and a managed, anti-free-market economic approach—as the norm for responsible, virtuous human life and political citizenship. In the wake of dwindling core curricula and declining emphasis on religious and transcendent educational purposes, sustainability is the new metanarrative. It shapes the curriculum, molds student life, and trains students to accept—even welcome and agitate for—the introduction of greater social control.

In all these ways, the college campus has become the center of this sustainability authoritarianism.

On campus, sustainability poses five particular problems. First, it displaces open inquiry into important subjects with a blatant appeal to authority, usually in the form of declaring that there is a "consensus" among scientists that forecloses the need for further examination of what is precisely in question. Second, it orients faculty research and teaching into a prescribed orthodoxy. Not only are certain questions shut out, but certain answers are locked in. It turns higher education into a form of indoctrination. Third, it subjects students to a regime of constant manipulation, or "nudging," to keep them psychologically attuned to the special demands of the ideology. Fourth, it undermines the ideals of liberal arts education, which require mindful attention to many matters that sustainability now deems irrelevant. For example, many colleges are embracing something called "the environmental humanities," which pushes aside the traditional humanities curriculum. And fifth, it diverts students into pointless battles at the expense of learning how to participate meaningfully in American civic life.

Hurdles

Critiquing the campus sustainability movement requires that we overcome several hurdles.

1. Who are we to question a movement that has such broad and deep support?

The National Association of Scholars is a membership organization founded in 1987 to "advance

---

22   Adam Corner, "Morality Is Missing from the Debate About Sustainable Behaviour," *Guardian*, July 19, 2013. http://www.theguardian.com/sustainable-business/social-justice-behaviour-climate-change

NAS

reasoned scholarship in a free society." We uphold standards of excellence in teaching and research, and more broadly, we advocate for the intellectual freedom without which such standards would be a dead letter. This has brought us into confrontation with those who favor using the university as a tool of political activism. Long before the sustainability movement appeared on the scene, the NAS was fighting the rise of illiberal ideologies that were bent on shutting down important debates in favor of giving students their one-sided views of history, philosophy, literature, law, and science.

Our critique of sustainability continues this long effort to call American higher education back to its higher and better principles. Rigorous academic standards require the pursuit of truth. To this end we support vigorous debate on important issues, where all the best arguments and all the pertinent evidence is put forward and every voice that abides by the rules of argument and evidence is welcome.

We first noticed the CSM, the campus sustainability movement, in 2008, and began to document it in a series of articles posted to the NAS website, and in other publications such as *The Chronicle of Higher Education*. In 2010, we published a special issue of our journal *Academic Questions* devoted to the sustainability movement. All told, before the publication of this report, NAS published some 200 articles on the CSM, which established NAS as the leading critic of the movement.

Our standing as a critic of the CSM has one other important component. NAS has a long history of approaching issues in higher education by way of in-depth studies. Our approach is to gather and analyze a large amount of detailed information. Our best known studies include *The Dissolution of General Education: 1914-1993; Losing the Big Picture: The Fragmentation of the English Major Since 1964; The Vanishing West: 1964-2010; Recasting History: A Study of U.S. History Courses at the University of Texas and Texas A&M University; Beach Books: What Do Colleges and Universities Want Students to Read Outside Class?* and most importantly, *What Does Bowdoin Teach? How a Contemporary Liberal Arts College Shapes Students.*

These reports go to great lengths to describe aspects of higher education that others take mostly for granted. The reward for our efforts is that we find important things that others have not noticed. In presenting this study of the campus sustainability movement, we likewise aim to take up a subject that might, at the outset, appear to be thoroughly familiar and ordinary. But by looking at it with care, we intend to bring it to life in a new way.

> 2. CSM is rooted in the claim that man-made global warming is an established scientific fact, no longer open to serious doubt. Does the critique of the campus sustainability movement require a robust rejection of the "global warming consensus?"

NAS

The National Association of Scholars takes no position on the existence of global warming, its magnitude, its causes, or possible remedies. These are, properly, matters of scientific inquiry. NAS's membership includes many scientists who have relevant expertise, but we are not organized as a body to pursue research on scientific questions and we claim no special authority on the answers.

This will pose a difficulty for some readers. The CSM takes massive man-made global warming as scientifically established and beyond the need of further inquiry. We reject that kind of certainty on principle. Science does not advance by declaring certain hypotheses to be beyond question. Clearly some theories are so robust that they are unlikely to be called into serious doubt anytime soon, but that does not apply to "climate science," which is a field of inquiry heavily dependent on computer models, speculative formulations, and reconstructed data. Moreover, thousands of scientists have skepticism about the so-called "scientific consensus" on global warming. As the atmospheric and space physicist S. Fred Singer puts it, "the models cannot reproduce the observations."[23]

In this light, we side with those who argue that there is no real scientific consensus and that large and important areas remain open for further inquiry. It could be that the results of that inquiry will vindicate the "consensus" version of global warming hypotheses. Or it could be that the results of the inquiry will discredit the "consensus" version of climate change.

Regardless of how that turns out, we see a need for a careful critique of the campus sustainability movement. That's for three reasons. First, CSM is built on unwarranted certainty about how the earth's atmosphere and oceans will respond to relatively minor increases in atmospheric carbon dioxide. Second, CSM misappropriates climate science as a source of authority for changes in public policy that have scant connection to climate. Third, CSM has become an ideology that undermines liberal education—and would undermine it *even if the global warming hypothesis turns out to be true*.

The CSM, of course, treats the kind of agnosticism we adopt as untenable. It adopts a position of "You are for us or against us." Can NAS act as the Switzerland of climate change? Our position is rooted in NAS's long history of standing for intellectual freedom on contentious issues. The best way to do that is to ensure that the best arguments from all sides are fairly represented.

The advocates of the theory of anthropogenic (man-made) global warming (AGW) have a long history of efforts to suppress the publication and expression of contrary views. By treating both sides of the debate as legitimate, we inevitably break this taboo and appear to range ourselves on the side of the skeptics.

23 S. Fred Singer, "The China Climate Accord: A Bad Deal for the US," *American Thinker*, December 8, 2014. http://www.americanthinker.com/articles/2014/12/the_china_climate_accord_a_bad_deal_for_the_us.html

NAS

Appearances notwithstanding, we remain off-shore and uncommitted on the question of the validity of AGW hypotheses. Where we place our stake in the ground is on the indispensable importance for higher education of healthy debate, which requires creating a situation where all parties have both a right and an opportunity to speak.

> 3. CSM employs a variety of techniques aimed at "nudging" students to adopt the views and practices the movement favors, rather than persuading students of the validity of these views. Is this a form of manipulation that deserves censure or simply an ordinary part of marketing new ideas?

We devote a chapter of this study to the phenomenon of colleges "nudging" students to align their ideas and habits with the dictates of the sustainability movement. In our view these practices are inappropriate. Our chapter on nudging is of particular value because it describes the everyday experience of students on college campuses in which the sustainability movement has achieved authoritative control.

NAS views the domination of college campuses by illiberal ideologies of any sort as harmful to the better purposes of higher education. But some ideologies, in our view, are worse than others. Sustainability has several components that we think mark it as especially detrimental to liberal education. It forecloses open inquiry on matters such as global warming. It fosters hostility to Western civilization, to free markets, and to personal freedoms. (We turn to these in due course.) With "nudging" we come to the importance of treating students as people who are being specifically educated to "think critically" and exercise independent judgment. The decision of a college to "nudge" rather than persuade sounds a note of disdain for the right of students to make up their own minds.

> 4. CSM supporters range from radical activists who issue apocalyptic warnings to pragmatic moderates who seek conventional reforms such as recycling. Is it fair to treat such a range of perspectives as a single movement?

The CSM spans views ranging from extremists who express a profoundly misanthropic view of humanity to moderates who simply want to take reasonable precautions and perhaps economize on energy expenses. The movement, in effect, extends from prophets of an apocalypse to apothecaries of global healing. Sometimes we see the same people in different moods, arguing apocalypse in the morning and apothecary in the afternoon.

To describe the campus sustainability movement as a whole, we need to present as fairly as possible all of the positions in this spectrum. But describing the full spectrum does not mean devoting equal space to every part of it. This report pays proportionally greater attention to the apocalyptic side, where the campus movement derives much of its force and devotes much of its time. Aggressive activism seems to require these scare tactics that emphasize runaway global warming that will drown coastal cities and plains, set off enormously destructive storms, drastically reduce the world's food supply, and in other ways let loose a train of catastrophes. Global warming isn't the only apocalyptic scenario evoked by the CSM, only the most common one. Other apocalyptic scenarios include catastrophes ensuing from genetically modified foods, nuclear war or nuclear accidents, overuse of the world's potable water, or the immiseration of humanity brought on by the unchecked growth of capitalism. These apocalyptic narratives are the backbone of the campus sustainability movement.

Our task in this report is to describe the campus sustainability movement, not to weigh these apocalyptic scenarios each on its merits. But because the threat of manmade global warming is conspicuous as the most widely enunciated reason for the movement, we recognize the need to present a basic account of the arguments for and against it. This is presented in Chapter 3.

As for the many supporters of milder and generally non-apocalyptic versions of the CSM, our task is to describe their outlook and concerns to the extent that these are distinct from the movement as a whole. This moderate wing of the movement includes scientists whose research is funded in large part because advocacy of the manmade global warming thesis has prompted public and private support for work on "climate change"; humanists who have reconfigured their own scholarly interests and teaching priorities to match the movement; and university facilities staff who have been rebranded as "sustainability" workers with or without actual changes in their responsibilities.

Colleges and universities as institutions naturally arrange themselves on the side of the movement that favors incremental action for long-term changes, rather than abrupt, utopian schemes such as an "ending capitalism" or eliminating altogether the use of fossil fuels. College and university presidents occasionally indulge in some of the more apocalyptic rhetoric of the CSM, but in the end these are institutions that are planning for a tomorrow in which children will pursue college degrees after high school rather than revert to subsistence horticulture or hunting and gathering.

We mention the return-to-the-primitive options not as caricatures of the sustainability movement but as matters that some of its leaders in all seriousness do advocate, and that have actually gained a small following among contemporary college graduates. Recently one of us spoke to the mother of a young woman who graduated in 2013 from Wesleyan College and who, out of her devotion to the ideals of sustainability, has since been living in a tent in Hawaii attempting to survive on the produce of her garden. Such cases are no doubt exceptional but they illustrate how this CSM movement can indeed reshape the minds and ambitions of young people away from mainstream goals to entanglement in eccentric fantasy.

The university may be fundamentally an instrument of the continuity of civilization but it also has had a byway that takes a certain number of students into psychological cul-de-sacs. Marxism was the cul-de-sac for many American students in the last century. Diversity became a cul-de-sac for many students in the last quarter-century. Sustainability is the new cul-de-sac for those who are susceptible to the lure of utopian ideology.

Ideology

Our critique of the sustainability movement makes significant use of the concept of "ideology." It is a word with a rich history, much of which is relevant to what follows. But we will leave that history in the background. By "ideology" we mean a doctrine that is self-contained. It presents itself to the believer as a body of premises that are self-evidently true and important. It anticipates arguments against those premises and has built-in reasons for the believer to reject such arguments. It also anticipates facts that will appear to be at odds with the premises and offers the believer built-in intellectual and emotional maneuvers to discredit such facts.

An ideology has both intellectual and emotional content. It asks for belief or assent, but it also demands action. It requires the believer to conform his life to the doctrine, to help recruit others to the belief, and to participate in some larger struggle to bring the world into alignment with the belief.

Finally, an ideology is always in conflict with something. It assumes the existence of people who have other beliefs, and it assumes that these other beliefs are invalid.

A well thought-out system for understanding the world is not entirely bad, of course. The main task of philosophy is to come to understand an integrated, complete picture of reality and to live well within it. Ideology, by contrast, fits the facts to doctrine and takes on some of the characteristics of a religion.

A number of observers, perhaps most significantly the French social theorist Pascal Bruckner, have written at length about the manner in which the sustainability movement has absorbed much of the dynamic of traditional Christianity, selectively picking up its themes of sin and condemnation in an increasinglysecular world. Sustainability, like Christianity, offers a view of the Earth as once-pristine and pure but now fallen; it recognizes the sinfulness of humanity; it offers forms of expiation and absolution; and it puts these elements together in a master narrative of an impending catastrophe that will punish mankind for its iniquity. Sustainability, like Christianity, insists on our guilt, which is both collective and individual. Sustainability also elevates certain individuals as exceptional and able to show others the way. The prophets or saints or the elect of sustainability are those who foresee the catastrophe ahead and are warning us now. Unlike Christianity, which sees redemption ultimately actualized in the next life, sustainability's adherents believe it is in their power to redeem themselves and other sinners here and now.

NAS

Some proponents, too, have embraced the religious label. The outgoing chair of the IPCC, Rajendra Pachauri, resigning after accusations of sexual harassment, commented in his letter of resignation that "For me the protection of Planet Earth, the survival of all species and sustainability of our ecosystems is more than a mission. It is my religion and my dharma."[24]

Whether sustainability is a religion or is merely like a religion is too fine a point for our purposes in this report. We will stay with the thesis that sustainability is an ideology with some religious overtones. This does not mean that everyone in the CSM has embraced the ideology with the same fervor or the same degree of resistance to counter-argument and discrepant evidence. We will deal with these matters as they come up.

Catastrophe

The sustainability movement is an ideology, but it is also a form of catastrophism. We have touched on this already, but it is worth adding that real and imagined catastrophes play a very special part in Western thought.

The Bible, of course, offers the narrative of the flood survived by Noah and his family. Plato in his work *Timaeus* transmitted the story of Atlantis, an island civilization that is destroyed when it sinks in the ocean. A key debate in the development of modern geology was whether the earth was shaped by colossal events or by gradual processes. The latter, called uniformitarianism, gained the upper hand in scientific theory for the better part of two centuries, partly because it fit so well with Darwinian theory of natural selection. In the uniformitarian picture, catastrophes of course do occur: volcanoes erupt, continental ice sheets form, etc. But such events can be seen as long-cycle parts of an underlying, generally uniform order. Uniformitarianism, however, began to lose some of its standing in the 1980s when the Alvarez hypothesis—that the extinction of the dinosaurs 65 million years ago was caused by a giant meteor impact—found support in the analysis of the Chicxulub crater in the Yucatan Peninsula.

Scientific catastrophism came back into vogue. Clearly, some major changes occur outside the range of even long-term cycles. It is best not to make too much of this change in the scientific zeitgeist, but it corresponds fairly well with the rise of global warming catastrophism. We happen to live in an age when the idea of abrupt and irreversible shifts in nature are seen as well within the range of possibility. The step from imagining that global catastrophes could happen to believing that one is happening right now deserves some consideration. One of the major alternatives to the manmade global warming hypothesis, in fact, is a version of the older uniformitarian theory. It holds that the rise in global temperature recorded

---

24 Letter from Rajendra Kumar Pachauri to Ban Ki-moon, Secretary General of the United Nations, February 24, 2015. http://www.ipcc.ch/pdf/ar5/150224_Patchy_letter.pdf

NAS

in the 20th century is a part of a regular long-term pattern associated with cycles of energy output by the sun.

Our contemporary receptivity to catastrophism, of course, has other sources as well. The fears of atomic warfare during the Cold War and the nightmare visions of environmental degradation that followed Rachel Carlson's 1962 book, *Silent Spring*, have primed us to imagine a world poised on the precipice of some kind of epic disaster. Catastrophism, once it becomes a habit of mind, becomes an expectation in search of something to be truly worried about.

Where Did Sustainability Come From?

Sustainability as a highly visible concept is usually dated to the release in 1987 of the United Nations' report, *Our Common Future*, which set out an idea that the nations of the world should seek a "sustainable development path." Sustainable development, according to the report, is:

> Development that meets the needs of the present without compromising the ability of future generations to meet their own needs.[25]

The commission that produced this report, headed by former Norwegian prime minister Gro Harlem Brundtland, was taking exception to an earlier UN action, the *Stockholm Declaration* of 1972, which had attempted to balance strong protections for the environment with the need for economic growth. The *Stockholm Declaration* enunciated 26 principles, among them Principle 11,

> The environmental policies of all States should enhance and not adversely affect the present or future development potential of developing countries, nor should they hamper the attainment of better living conditions for all. [26]

The Brundtland report rejected this development-to-enhance-living-conditions principle in favor of an approach that focused on development as a tool of improving the environment. The Brundtland commission also linked efforts to improve the physical environment with efforts to improve political and social conditions around the world.

Sustainability was thus launched in 1987 as a composite program—something that combined environmental goals with economic, social, and political objectives, and laid this encompassing project out as a multi-generational program and as a worldwide project. But global warming was not a major theme of the Brundtland report. The commission did indeed mention global warming in the report—fourteen times. These mentions, however, are in the context of listing many possible threats, including ozone   depletion,

---

25   World Commission on Environment and Development, *Our Common Future*. United Nations, 1987.

26   *Declaration of the United Nations Conference on the Human Environment*, United Nations Conference on the Human Environment, Stockholm, June 1972. http://www.unep.org/Documents.Multilingual/Default.asp?documentid=97&articleid=1503

NAS

nuclear fallout, acid rain, dryland degradation, hazardous chemicals, population movements, and species loss. Global warming is named, for example, as something that "may cause the flooding of important coastal production areas" within "40-70 years."[27] The earliest date that the commission imagined for the arrival of global warming was "by the 2020s."[28]

Today, of course, the sustainability movement is closely identified with the anthropogenic global warming hypothesis. It is important to understand how these two ideas became intertwined—a history that is included in Chapter 3. Global warming brought on by increases of a few molecules of CO_2 per million per year from the burning of fossil fuels may happen as well, and it may be a more plausible peril than some of the others.[29] Even if it is, the evidence so far is inconclusive.

The evidence for a different peril, however, is much more solid—the peril of our doing severe damage to our economy, to our personal freedom, and to our form of representative government by ceding our better judgment to an ideology. In the pages that follow, we will trace how that happened and suggest some ways to put that peril in its place.

The Report

The campus sustainability movement has become part of almost everything colleges and universities now do. Our aim in this report, however, is selective. We highlight the aspects of higher education that seem seriously harmed by the movement, and we fill in needed context as we proceed. Chapter 1 deals with the ways in which college curricula have been reshaped by the movement. Chapter 2 offers an anatomy of the campus movement: how it is organized both locally and nationally, who leads it, and how it goes about its business. Chapter 3 pulls back to the broader debate about global warming. In keeping with our agnostic position, we present both sides of the debate. Chapter 4 offers an in-depth look at how the movement attempts to motivate and sometimes manipulate students into cooperating with and, ideally, actively supporting the cause. Chapter 5 is our attempt to estimate the financial costs of the movement to

---

27   *Our Common Future*, pg. 106.

28   *Our Common Future*, pg. 146.

29   Observations from the Mauna Loa Observatory in Hawaii made by the Scripps Institution of Oceanography and the National Oceanic and Atmospheric Administration show the annual rate of increase in CO2 in the atmosphere to be at or below 2 parts per million.

Decade	Total Increase	Annual Rate of Increase
2004 – 2013	20.71 ppm	2.07 ppm per year
1994 – 2003	18.70 ppm	1.87 ppm per year
1984 – 1993	14.04 ppm	1.40 ppm per year
1974 – 1983	13.35 ppm	1.34 ppm per year
1964 – 1973	10.69 ppm	1.07 ppm per year
1960 – 1963	3.02 ppm	0.75 ppm per year (4 years only)

Pieter Tans, *Monthly Mean Concentrations at the Mauna Loa Observatory (PPM)*, National Oceanic and Atmospheric Administration, March 1958 - August 2014. http://co2now.org/images/stories/data/co2-mlo-monthly-noaa-esrl.pdf

NAS

a sample college. This is especially important because the movement often portrays itself as eventuating in major savings for the institutions that participate. In fact, it appears to be a cost-driver, not a form of thrift. Chapter 6 takes up the effort by sustainability activists to convince their colleges to divest their holdings in fossil fuels. We conclude the report with a set of recommendations and a set of appendices that introduce the voices of others who are playing active roles on our topic: a faculty member, a student, a scientist, and a lawyer.

CHAPTER 1: THE GREENING OF THE CURRICULUM

In "The Ethics of Eating," Professors William Starr and Andrew Chignell help Cornell students chew over "the questions" raised by breakfast, lunch, and dinner. One of their students, Lauren Thiersch, a sophomore in the School of Hotel Administration, summed up:

> This class demands one of two things: 1. That you defend the way you eat, or 2. that you change it. And in early February I stopped eating meat because of what I've read, watched, and learned in this class.[30]

The course was a four-credit philosophy class, but what Thiersch learned, judging by the syllabus, was a good deal of one-sided alimentary advocacy. "The Ethics of Eating" booklist is fat with organic memoirs and anti-factory farm diatribes seasoned only lightly with actual ethics. The syllabus includes Jonathan Safran Foer's memoir of going vegetarian, *Eating Animals*, and its counterpart, *Reclamation: A Tale of Blood, Betrayal, and Bioregional Meat* by ex-vegan turned "naturalist-hunter" Brad Dingman; *The Omnivore's Dilemma* by foodie-activist-journalist Michael Pollan; and *Food for Thought: The Debate Over Eating Meat*, edited by the animal ethicist Steve F. Sapontzis.

Students also watched the documentaries *Food, Inc.* (narrated in part by Michael Pollan) and *Our Daily Bread*, which portray ghastly images of factory farms and slaughterhouses as normal parts of corporate food production. At one point the class took a trip to a local meat processing plant to observe how slaughterhouses operate. Starr and Chignell also brought in two guest speakers during the semester: the vegan author Jonathan Balcome, who researches animal experiences of pleasure and is writing a book about the inner lives of fish, and Brian Wansink, the expert in eating behavior who got McDonald's to swap apple slices for French fries in Happy Meals and convinced snack companies to create 100-calorie single-serve portions as a way to curb mindless eating.

Vegetarianism may be a perfectly reasonable personal decision, but it departs substantially from the traditional list of educational purposes: sharpening students' faculties of reasoning and expanding their understanding of reality. Starr and Chignell aver that "The goal of this course is not to teach some preferred set of answers to these questions" but to give students "the basic tools" they need in order to "reflect clearly and effectively on the questions themselves."[31] But the lopsided booklist and speaker list, combined with the suggestion that most of our meals pose moral dilemmas and the requirement that students either "defend" or change their eating habits, offer pretty clear indications of which side Starr and

30 Lauren Thiersch, "PHIL 1440: Ethics of Eating," bigREDefined, April 18, 2014. http://blogs.cornell.edu/lauren/2014/04/18/phil-2411-ethics-of-eating/

31 PHIL 1440 – Ethics of Eating, Cornell Catalog 2013-2014. http://courses.cornell.edu/preview_course.php?catoid=22&coid=343914

NAS

Chignell hope their students come down on.

"The Ethics of Eating" is one of 403 courses that Cornell has categorized as sustainability courses.[32] Using the official guidelines delineated by the Association for the Advancement of Sustainability in Higher Education (AASHE) in a 300-page manual, Cornell distinguishes between classes that are sustainability-related and sustainability-focused:

- Sustainability-focused courses concentrate on the concept of sustainability, including its social, economic, and environmental dimensions, or examine an issue or topic using sustainability as a lens.

- Sustainability-related courses incorporate sustainability content as a course component or module.[33]

Cornell's sustainability-focused and related courses include the alarm-sounding "Earthquake!" (exclamation point included!),[34] the far-sweeping "Microbes, the Earth, and Everything,"[35] and the victim-fixated "Race & Social Entrepreneurship, Environmental Justice and Urban Reform."[36] "Magnifying Small Spaces Studio" teaches students how best to live in mini-spaces and answers the question, "In reducing one's carbon footprint, how small is too small?"[37] And Cornell offers a whole buffet of courses for those hungry for more of what Starr and Chignell served up. "Climate Change and the Future of Food,"[38] for instance, prepares students for the possibility of an agricultural collapse, while "Food, Farming, and Personal Beliefs"[39] compares the "personal value systems of farmers and consumers" and the relationships between religious faith and sustainability.

How does such an eclectic group of courses become part of the college curriculum at a major university? The idea, according to the David R. Atkinson Center for a Sustainable Future at Cornell, is that sustainability operates as an interpretive key that rises above the academic subject divisions to pick and choose pieces

---

32 Cornell Sustainability Courses, David R. Atkinson Center for a Sustainable Future, Cornell University. http://www.acsf.cornell.edu/education/curricula/

33 Version 1.2 Technical Manual, AASHE STARS, February 2012, page 39.

34 EAS 1220 - Earthquake! Cornell University Catalog, 2014-2015. http://courses.cornell.edu/preview_course.php?catoid=22&coid=337177

35 CSS 1120 - Microbes, the Earth, and Everything, Cornell University Catalog, 2014-2015. http://courses.cornell.edu/preview_course.php?catoid=22&coid=336931

36 ASRC 4330 - Race & Social Entrepreneurship, Environmental Justice and Urban Reform, Cornell University Catalog, 2014-2015. http://courses.cornell.edu/preview_course.php?catoid=22&coid=344667

37 DEA 2201 - Magnifying Small Spaces Studio, Cornell University Catalog, 2014-2015. http://courses.cornell.edu/preview_course.php?catoid=22&coid=337061

38 HORT 3600 - [Climate Change and the Future of Food], Cornell University Catalog, 2014-2015. http://courses.cornell.edu/preview_course.php?catoid=22&coid=344169

39 CSS 4910 - Food, Farming, and Personal Beliefs. Cornell University Catalog, 2014-2015. http://courses.cornell.edu/preview_course.php?catoid=22&coid=336958

from among them to weave back together in new, sustainable ways: It "transcends individual disciplines" but rests on "on a foundation of disciplinary understanding."[40] Hence it can wedge its way into all kinds of courses.

Degrees of Sustainability

Courses such as Cornell's "Ethics of Eating" or "Microbes, the Earth, and Everything" are quickly becoming staples of today's undergraduate student's educational diet. It is common now to find courses coded as "sustainability-relevant" in college catalogues. Often sustainability gets its own department.

According to the Association for the Advancement of Sustainability in Higher Education, 475 college campuses in 65 states or provinces offer a total of 1,438 academic sustainability programs, ranging from certificates to undergraduate degrees to master's and doctorate degrees.41 In the U.S. alone, there are 1,274 programs, representing all fifty states. That's in addition to the hundreds of institutions that offer freestanding elective sustainability classes. Sustainability has graduated from a hobby research interest to a full-scale academic "discipline" that undergraduates can major in, graduate students can specialize in, and professors can become experts in.

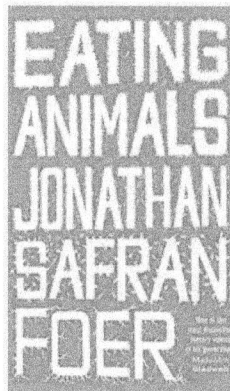

Eating Animals
by Jonathan Safran Foer

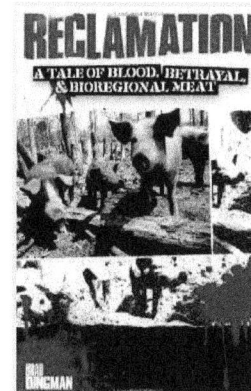

Reclamation: A Tale of Blood, Betrayal, and Bioregional Meat
by Brad Dingman

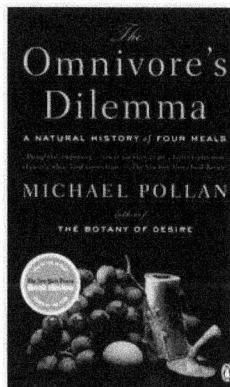

The Omnivore's Dilemma
by Michael Pollan

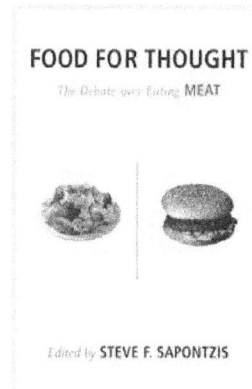

Food For Thought: The Debate Over Eating Meat
edited by Steve F. Sapontzis

The "disciplinary" status of sustainability is a bit open to question, though. It isn't a distinct science, like biology or physics; it isn't a distinct branch of the humanities, like philosophy or history; it isn't a social science, like economics or sociology. Rather it is a swirl of ideas and commitments that touch many things.

---

40   Cornell Sustainability Courses. http://www.acsf.cornell.edu/education/curricula/

41   AASHE Academic Programs Database. Association for the Advancement of Sustainability in Higher Education, as of February 23, 2015. http://www.aashe.org/resources/academic-programs/

But discipline or no, students can get a degree in it. The nation's first school of sustainability opened at Arizona State University in 2006. Housed within the Julie Ann Wrigley Global Institute of Sustainability, the school offers both bachelor of arts and bachelor of science degrees in sustainability, as well as minors in sustainability. Graduate students may earn a Ph.D., M.A., or M.S. in sustainability, a master's in sustainable solutions, or an executive master's for sustainability leadership.[42] Its dean of sustainability, Christopher Boone, has a Ph.D. in geography and is an expert in urban infrastructure and environmental justice.

Some colleges are devoted conspicuously to the teaching of sustainability. College of the Atlantic in Maine directs all students to focus on "human ecology."[43] Florida Gulf Coast University's motto is "sustainability, excellence, service"; its guiding principles commit professors to "instilling in students an environmental consciousness that balances their economic and social aspirations with the imperative for ecological sustainability."[44]

Unity College, whose slogan is "America's environmental college," focuses all of its course offerings around a core "Environmental Citizen curriculum" that every student must take.[45] The three expected learning outcomes include students displaying "dedication to sustainability," transforming into "engaged citizens and leaders who welcome diversity, work well with others, respect tradition and differing points of view, and help encourage a productive, communal way of life," and developing "an extensive knowledge of the sciences, social sciences, and humanities" and how these disciplines "connect to and inform environmental issues."[46] The college's mission statement commits Unity to providing "a liberal arts education that emphasizes the environment and natural resources" and to teaching each subject "through the framework of sustainability science."[47] In the midst of a national decline in core curricula, Unity College's decision to make sustainability courses a graduation requirement is telling.

Four hundred twenty-four of the sustainability programs on AASHE's list are baccalaureate programs leading to degrees in topics such as "Environmental Sustainability" (Rochester Institute of Technology) or "Sustainability Practice" (Lipscomb University). The majors range from earthy ("Sustainable Landscape Horticulture," University of Vermont) to artsy ("Architecture and Sustainability," Ferris State University), from business-focused ("Sustainable Management and Policy," Purdue University) to socially-minded

42 Degree programs, School of Sustainability, Arizona State University. https://schoolofsustainability.asu.edu/about/welcome-introduction.php

43 "About COA," College of the Atlantic. http://www.coa.edu/about-coa.htm

44 "Vision, Mission, & Guiding Principles," Florida Gulf Coast University, June 18, 1996. http://www.fgcu.edu/info/mission.asp

45 "Environmental Citizen Curriculum," Academics, Unity College. http://www.unity.edu/academics/a-distinctive-approach/environmental-citizen-curriculum

46 *Ibid.*

47 "Our Mission," Unity College. http://www.unity.edu/mission

NAS

("Sustainable Community Development," Northland College). There are energy-focused programs ("Energy and Sustainability Policy," Pennsylvania State University), global-warming inspired themes ("Global Environmental Change and Sustainability," Johns Hopkins University), options for aspiring teachers ("Environmental Education," Unity College), and economics approaches for the analytical types ("Environment, Economics, and Politics," Claremont McKenna College).[48]

Figure 1. Map of Sustainability Degree Programs[49]

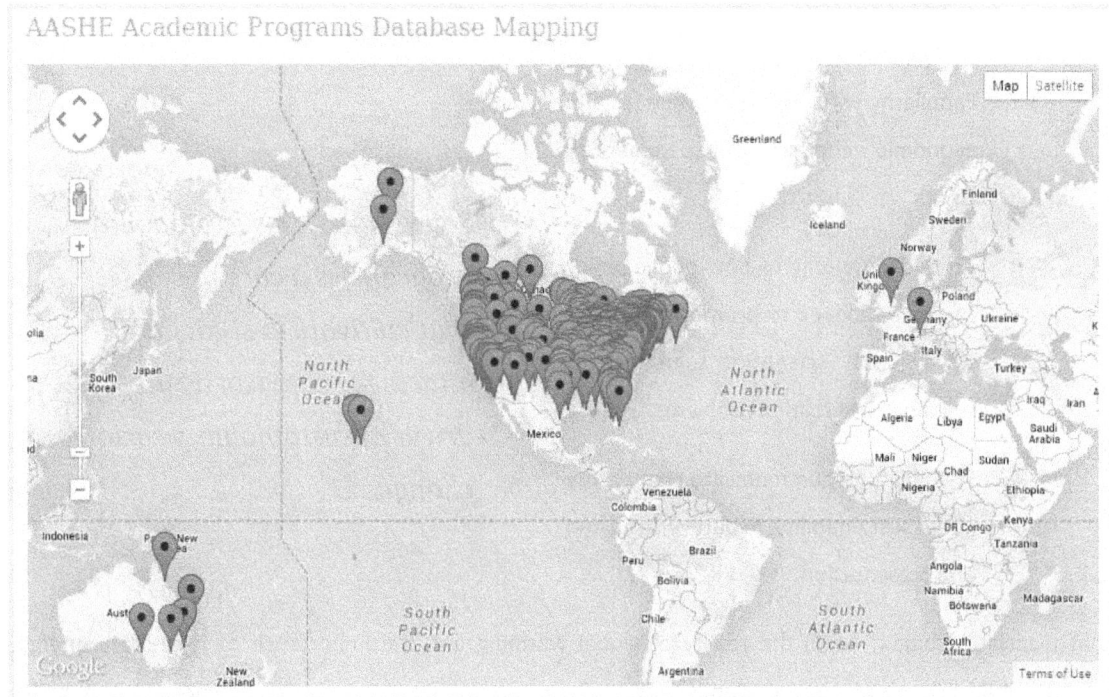

What do students learn in these programs, besides an awareness of how expansive sustainability can be? Something like what they learn in Cornell's "Ethics of Eating"—a little science, a bit of economics, a dab about ethics, and a great deal of social theory and political advocacy. Sustainability is less like a newly discovered, previously missing layer on the educational food pyramid and a more like a casserole. It borrows bits and scraps from here and there and, under the heat of the oven, bakes them into some semblance of a cohesive, tasty entrée. It can offer courses in ethical eating and environmental poetry at the same time that it offers a few in trash studies and sociology. There's something there for everyone.

---

48   Sustainability-Focused Baccalaureate Degree Programs, Association for the Advancement of Sustainability in Higher Education. As of September 11, 2014. http://www.aashe.org/resources/academic-programs/type/bacc/

49   AASHE Academic Programs Mapping, Association for the Advancement of Sustainability in Higher Education. http://www. aashe.org/resources/academic-programs/map/

NAS

That salmagundi nature is redolent at the University of South Dakota (USD), where undergraduates may ingest sustainability in the form of either a Bachelor of Science or a Bachelor of Arts degree in sustainability, depending on whether they wish to specialize in the natural sciences or the social sciences. Those with less hearty palates may also select a minor in sustainability to complement whichever main course of study they choose. In all three versions, students aim to achieve four wide-ranging learning goals by the time they complete their programs:

1. An understanding of the fundamental scientific concepts that contribute to assessing the sustainability of human activities (e.g., environmental impact and resource depletion) and to evaluating sustainable technologies (e.g., energy and food production).

2. Familiarity with the social, political, and economic context of sustainability issues.

3. An understanding of how public policy can be employed to promote or inhibit social and scientific solutions to sustainability-related problems.

4. The ability to communicate proficiently about sustainability, in both written and oral presentation.[50]

The pièce de résistance at South Dakota's sustainability programs is a core of classes that cultivate in students a taste for activism as the driver of sustainability social change.

That translates roughly to 1) the reality of global warming and the importance of renewable energy, (despite the presence of real academic debates over the merits of the science behind both); 2) belief in the harm of traditional social and economic structures and the need for reimagining society along progressive political lines; 3) the need for mass environmental activism as the solution to political problems; and 4) the ability to persuade skeptics and dissenters to conform with standard environmental thought, or, failing that, to keep silent.

The pièce de résistance at South Dakota's sustainability programs is a core of classes that cultivate in students a taste for activism as the driver of sustainability social change. Students must take one course in public policy ("Introduction to Public Policy") where they learn about the "dynamics of agenda setting, policy formulation, implementation, and evaluation."[51] They also take three courses in sustainability: "Sustainability and Society," "Sustainability and Science," and "Sustainability Capstone." "Sustainability

---

50   College of Arts and Sciences: Sustainability, University of South Dakota. http://www.usd.edu/arts-and-sciences/sustainability/undergraduate.cfm

51   "Introduction to Public Policy," University of South Dakota Catalog 2014-2015. http://catalog.usd.edu/preview_program.php?catoid=15&poid=2413#

and Society" assesses sustainability as a "framework" for addressing "complex societal issues," which according to the university include "food systems, social justice, and sustainable development."[52] Welfare reform, drug abuse, and bioethics concerns are notably missing from the list.

From there, the curriculum moves into a smorgasbord of environmental science and economics, field experiences and internships, along with ecology and climate science for the natural sciences concentration, and population studies, food studies, and communications for the social sciences concentration.

What exactly "sustainability" is as a discipline is hard to pin down. It appears to be a particular way of interpreting reality rather than a particular portion of knowledge to study. In that regard, sustainability is a bit like "multicultural studies" or "gender studies": a broad range of separate interests and subjects that mesh together only by the interpretive lens through which the student looks. In the case of sustainability, that lens is something along the lines of a deep-seated fear of depletion and unequal distribution of resources in the three spheres of the environment, the economy, and society, rectified by the solutions of collective political action and individual frugality.

Integrated

Increasingly sustainability is not just a subject a student may opt to study if he wishes, but an inescapable, automatic part of all disciplines and subjects. Sustainability spreads outside its disciplinary silo and into the curriculum at large as a set of assumptions that color even non-environmental courses. Students in sociology or business programs—or hotel administration, as in the case of Cornell's Lauren Thiersch—find themes of sustainability in their non-sustainability electives, or in their (non-sustainability) core major requirements.

"Integrated" is the term colleges and universities use when they speak of fitting sustainability into the full spectrum of their academic offerings. Cornell in its 2013 Climate Action Plan, which sets out the university's long-term sustainability goals, speaks of "integrating sustainability into students' educational experience"[53] and developing a sustainability educational program that will be "integrated with freshman orientation, undergraduate club leadership development, residential life, and professional development trainings."[54] Cornell is only one of hundreds of universities to speak in this manner. "Integration" has a long sustainability pedigree. The text of the American College and University Presidents' Climate Commitment (launched in 2006) draws on the term when it notes,

---

52 "Sustainability and Society," University of South Dakota Catalog 2014-2015. http://catalog.usd.edu/preview_program.php?catoid=15&poid=2413#

53 "2013 Climate Action Plan Update & Roadmap 2014-2015," Cornell University, 2013, pg. 9. http://rs.acupcc.org/site_media/uploads/cap/167-cap_1.pdf

54 *Ibid* pg. 29.

NAS

Campuses that address the climate challenge by reducing global warming emissions and by integrating sustainability into their curriculum will better serve their students and meet their social mandate to help create a thriving, ethical and civil society.[55]

"Integrated" sustainability might at first involve the mere introduction of a few sustainability-themed courses alongside the other, regular course offerings, but it soon aims to mean something more. Advocates want sustainability to become incorporated into the curriculum much in the way yeast permeates bread, and not as nuggets of chocolate, sprinkled here and there, flavor cookies. The one spreads everywhere, actively changing the substance of the entire dough, and goes subtly unnoticed except when absent. The other offers rich, concentrated flavor, but only in a few select bites.

Second Nature senior fellow Peter Bardaglio made the goal of complete saturation of the curriculum clear in his 2007 manifesto, "'A Moment of Grace': Integrating Sustainability into the Undergraduate Curriculum." Writing for the journal *Planning for Higher Education*, Bardaglio argued that sustainability advocates had a powerful, but narrow sliver of time in which they could radically remake the college experience with sustainability as the foundation. "The full integration of sustainability into the curriculum poses a fundamental challenge to the dominant paradigm in higher education," Bardaglio wrote. One collective, concentrated push to settle sustainability comfortably at home within all academic disciplines could firmly establish sustainability in campus values and priorities.[56]

Bonanza

Bardaglio spotlighted four colleges where a few activist faculty members armed with a budget, a winsome spirit, and determination had succeeded in treating sustainability as a multidisciplinary endeavor. Professors at Northern Arizona University, Emory University, Berea College, and Ithaca College had each created campus centers of some kind to serve as hubs for sustainability education.

Peter Bardaglio

Northern Arizona's program was the first and most influential of the four. The locus of sustainability curricular change at the university was a small campus center known as the Ponderosa Project. The name hinted at the scope of its ambition. "Ponderosa" was not only the name of the majestic pines that towered above the university campus, but also the fictional gigantic ranch that was the setting for the 1960s television Western *Bonanza*.

---

55   Text of the American College and University Presidents' Climate Commitment. http://acupcc.org/about/commitment

56   Peter Bardaglio, "A Moment of Grace," *Planning for Higher Education*, 2007, pg. 17.

In 1992, just as Second Nature was launching its first campus outreaches, Northern Arizona's director of English composition Geoffrey Chase and education professor Paul Rowland partnered with Anthony Cortese (then president of Second Nature) to figure out a way to bring sustainability into college campuses. At a Second Nature workshop, Rowland and Chase helped develop the idea of recruiting other faculty to teach sustainability-themed courses, or at least to incorporate examples, readings, or other material in their classrooms.

Three years later, having secured a grant from the U.S. Department of Energy, Rowland and Chase gathered twenty Northern Arizona faculty members at a two-day workshop, paid them each a $1,000 stipend, and coached them in revising their syllabi to incorporate sustainability. The workshop became an annual event, and by 2007, when Bardaglio conducted his survey of the state of sustainability education, the Ponderosa Project had led to the revision of 262 undergraduate classes and 97 graduate classes in 31 different departments.[57]

The kinds of revisions aimed for involve making sustainability part of the standard educationalexperiences of the average student at Northern Arizona University, rather than a segmented discipline that thestudent had to seek out and choose. The goals of the project, listed on the Ponderosa Project website, emphasize the importance of exposing all students from all disciplines to the principles of sustainability:

- "Green the Curriculum" so that the theme of environmental sustainability is introduced and reinforced throughout students' educational experiences

- Educate students in all courses of study about the implications of environmental sustainability in their chosen careers[58]

That spirit became a benchmark for sustainability education around the country. Bardaglio summed up the vision of the Ponderosa Project that spread to similar programs at other campuses:

> *Central to the Ponderosa Project has been the belief that the entire university, not just a single program, is responsible for sustainability. Project leaders have insisted that the best way to educate students about sustainability is to integrate it into a variety of subjects, rather than "ghettoize" it in an environmental studies program.*[59]

Here is perhaps where sustainability becomes most powerful, hidden in courses where the unsuspecting student meets it not as a tenet to be discussed and investigated, but a baseline assumption on which all subsequent scholarship and dialogue rests. The average student, if he has not previously made up his

---

57 Bardaglio, pg.18.

58 "Goals," Ponderosa Project. http://www2.nau.edu/~ponder-p/index.html#philosophy

59 Bardaglio, pg. 19.

mind on sustainability, or does not guard against the assumptions he encounters in class, almost cannot help being formed into an adherent of sustainability.

The Piedmont Project

Emory University offers another glimpse at what non-"ghettoized" sustainability education looks like. At the Piedmont Project there (one of the Ponderosa Project spin-offs that Bardaglio found so encouraging), environmental advocate Peggy Barlett took a survey of participating faculty members to find out in what manner they had fit sustainability into their courses. Barlett, the Goodrich C. White Professor of Anthropology at Emory and one of the leaders of the Piedmont Project, had led workshops annually at Emory and at other universities, where she trained professors in ways to teach their students sustainability alongside their primary professional disciplines.

Barlett found that "the vast majority" of Piedmont participants had changed their pedagogy to include experiential learning, new outdoor exercises, or new ways of engaging students. Meanwhile, 44 percent revised their courses by adding new labs, homework, or research projects; 64 percent developed a new unit or module; and 34 percent completely reoriented their course with a new paradigm.[60]

Exactly what this "new paradigm" represents, Barlett and her colleagues at the Piedmont Project don't quite define. But other sustainability leaders offer a hint of what it might entail. Dickinson College President Neil B. Weissman explains in "Sustainability & Liberal Education: Partners by Nature," that sustainability provides an educational metanarrative that "powerfully validates the liberal arts" and that glues together the diverging academic disciplines. That metanarrative cuts down "disciplinary silos" and offers "holistic systems thinking, the ability to make connections, interdisciplinarity, and 'lateral rigor'"—characteristics that Weissman found especially attractive when he launched Dickinson's own version of the Ponderosa Project, the "Valley and Ridge Education for Sustainability" group.[61]

Integrated sustainability education, then, is not merely a tactic to reduce campus water and energy usage, or to help the college earn a greener reputation, or even to train students to shrink their environmental footprints—though it does involve, to varying degrees, all of these. Instead, sustainability becomes the overarching purpose of education itself, a pedagogic goal broad enough to speak to "virtually all academic disciplines" but substantive enough to demand that the disciplines "enter into dialogue."[62] Sustainability offered an underlying foundation, the overarching *telos*, and the intermediary substance of education all at once.

---

60 "About the Piedmont Project," Emory University. http://piedmont.emory.edu/About.html

61 Neil B. Weissman, "Sustainability & Liberal Education: Partners by Nature," *Liberal Education*, Fall 2012, Vol. 98 No. 4, pg. 2.

62 Weissman, pg. 2.

Weissman was not the first to make this point. Former Cornell President Frank H.T. Rhodes argued a similar case in a 2006 *Chronicle of Higher Education* op-ed titled "Sustainability: The Ultimate Liberal Art." Rhodes's piece, published shortly after the initial launch of the Presidents' Climate Commitment, encouraged colleges and universities to adopt sustainability as a central educational goal because it fit the ancient liberal arts tradition of preparing

> *Sustainability becomes the overarching purpose of education itself, a pedagogic goal broad enough to speak to all academic disciplines.*

students "for citizenship, for participation in a free society" and thus provided a "new foundation for the liberal arts and sciences."[63] Rhodes described what this liberal arts education built on the foundation of sustainability would look like:

> *What might such a foundation entail? Certainly some significant exposure to the appropriate sciences: geology, natural resources, ecology, and climatology. Certainly, too, some understanding of social interaction sociology, economics, and history. And also, surely, some extensive familiarity with the great issues and themes of human inquiry, self-reflection, and moral consideration that have guided human conduct and reflected human creativity — with the arts and the humanities, in other words. And to anchor everything in the present, some review of the practical arts of technical discovery and invention, especially in relation to the broad issues now confronting us.*

Minus the climatology and sociology, Rhodes's list looks much like a photocopy of a traditional liberal arts curriculum. "But, in fact, it would be different," Rhodes claimed, not in its choice of subject matter but in "the new focus, added coherence, and stark immediacy that it (sustainability) would provide." Rhodes concluded confidently, "Sustainability, after all, is the ultimate liberal art (and science)."

Environmental Humanities

The ideas of sustainability as the "ultimate liberal art" and as an overarching metanarrative for human existence have led to the creation of some new hybrid disciplines that are neither individual sustainability-tinged electives nor narrowly focused sustainability programs, but distinct branches of larger disciplines that then adopt a special focus on matters of sustainability.

One of the most prominent of these disciplines is a new field calling itself the "Environmental Humanities," a program beginning to pop up at institutions across the country. Universities offer undergraduate and graduate degrees in the field, national and international environmental humanities societies and

---

63   Frank H.T. Rhodes, "Sustainability: The Ultimate Liberal Art," *Chronicle of Higher Education*, October 20, 2006. http://chronicle.com/article/Sustainability-the-Ultimate/29514/

NAS

associations have formed, and a rising number of journals investigate relationships between nature and society.

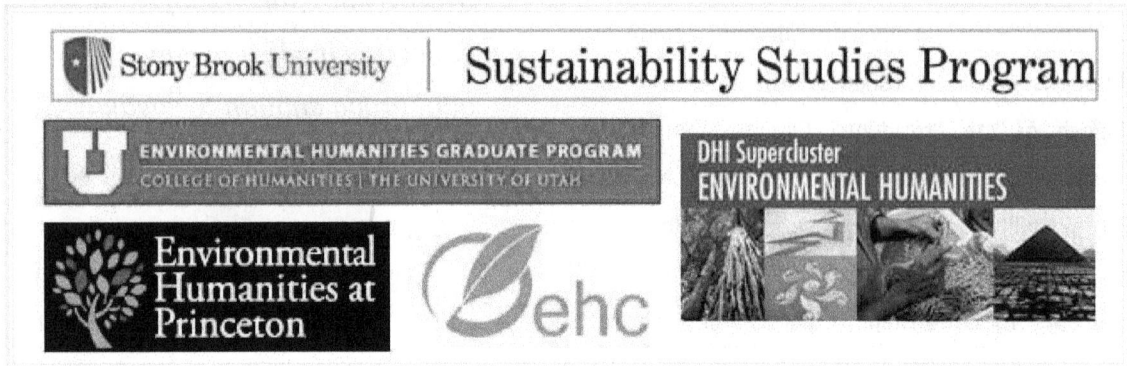

Environmental Humanities programs of study and research centers are sprouting at places such as Stony Brook University, University of Utah, University of California-Davis, Princeton University, and University of California-Santa Barabara.

The field hopes to marry the arts and humanities with the environmental sciences. That partly means taking a historical, anthropological look at how humans have collectively treated and theorized about nature—reading how they anthropomorphized or deified nature in literature, examining how animals and plants are depicted in artwork, and understanding the cultural values and norms embodied in poetry, song, and dance. It partly means making fields such as English, art, and rhetoric serve as conduits for disseminating environmentalist messages and scientific findings to a lay audience and for advocating social and legal action on behalf of the physical environment.

It also means exploring the possibilities of what climate change will mean for human civilization's existence and character. Bowdoin College English professor David A. Collings explores this third theme in his book *Stolen Future, Broken Present: The Human Significance of Climate Change*, published by Open Humanities Press as part of its "Critical Climate Change" series. Collings writes that climate change is poised to imminently endanger civilization, and while all hope is not doomed yet, "it is time for us…to contemplate, for the first time, what it means for us if we fail" to avert global warming.[64] Now that science has shown us global warming, and technology has so far failed to solve environmental problems, "we now face questions not simply about the scientific, technological, economic, or political dimensions of this crisis, although they remain crucial, but also about its *human* significance."[65]

These three themes show clearly at the Environmental Humanities major at Stony Brook University, one of a dozen or so universities with such programs. (Some of the other institutions include the University

---

64   David A. Collings, *Stolen Future, Broken Present: The Human Significance of Climate Change*, University of Michigan Library: Open Humanities Press, 2014, pg. 13.

of Vermont, the University of Utah, and the University of Oregon.) The historical perspective approach is evident in "Beyond Eden: Contact Narratives, Origins, and Sin," in which students research how five hundred year-old Pueblo, African, Spanish, British, and Shinnecock literature shapes contemporary themes about nature, human origins, and sins,[66] or in "Native American Texts and Contexts," a look at the American Indian oral tradition, poetry, history, and other writings.[67] The rhetorical and political training for advocacy makes an appearance in classes such as "Environmental Writing and the Media," or "Collective Action and Sustainability."[68] The concern for climate change's effects on humans is evident in "Civilizations and Collapse," which presents case studies in how human groups have in the past reacted to environmental changes,[69] and "The Maya," an ethnographic course that pays "special attention" to "the ways in which environmental and agrarian issues have impacted this diverse group of   peoples."[70]

All told, it's a hodgepodge of science courses mixed with boutique courses that sound, by turns, a little nouveau humanities, a little identity studies, and a little social science lite. Undergraduate Environmental Humanities majors must take courses in cultural anthropology, "Ecoaesthetics in Art," "Mathematical Thinking," "Introduction to Sustainability," and a choice of two from a list of five science courses: physical geography, "Chemistry, Environmentand Life," oceanography, "Organismsto Ecosystems," and "Introduction to the Natural History of Long Island." Students also have to take three one-credit "Career and Leadership Skills" courses and a three-credit "Integrative, Collaborative Systems Project."[71] They also take seven courses from a wide range of approved courses (reaching from "Extreme Events" to "Peoples and Cultures of South America" to "Theory and Design of Human   Settlement").

But the Stony Brook program, along with its counterpart programs at other universities, is motivated by more than a vague inkling for more interdisciplinary study or a hunch that the humanities might be indirectly useful to the cause of environmentalist advocacy. More than any other discipline, the Environmental Humanities portray the pinnacle of sustainability's influence as an interpretive framework for viewing all of human and natural reality.

---

66   "EHM 310: Beyond Eden: Contact Narratives, Origins and Sin," Stony Brook University Course Bulletin, Fall 2014. http://sb.cc. stonybrook.edu/bulletin/current/courses/index.pdf

67   "EGL 379 - J: Native American Texts and Contexts," Stony Brook University Course Bulletin, Fall 2014. http://sb.cc.stonybrook.edu/bulletin/2014/spring/courses/egl/#379

68   Environmental Humanities Major, Stony Brook University, Fall 2014. http://sb.cc.stonybrook.edu/bulletin/current/academicprograms/ehm/index.pdf

69   "EHM 314: Civilizations and Collapse," Stony Brook University Course Bulletin, Fall 2014. http://sb.cc.stonybrook.edu/bulletin/current/courses/index.pdf

70   "EHM 386: The Maya," Stony Brook University Course Bulletin, Fall 2014. http://sb.cc.stonybrook.edu/bulletin/current/courses/index.pdf

71   Environmental Humanities Major, Stony Brook University, Fall 2014.
65   *Ibid.*

George Handley, professor of humanities at Brigham Young University and one of the best-known proponents of the environmental humanities as an academic discipline, explains that "environmental humanities is no mere thematic approach to the study of culture, a sort of tree-hugger's tour of the great works of civilization," as if it saw nature as a mere subject matter for writing, singing, dancing, and philosophizing about. He rejects the idea that the discipline might "involve landscape painting but not necessarily the broader field of art, or it might include nature writing and nature poetry but not necessarily novels set in an urban context."[72] Instead,

> If we take seriously the challenge posed to human culture by the question of the natural world, we begin to see that there is little or no room to insist that "nature" and "culture" occupy separate and distinct arenas of our experience. And if this is the case, either all of nature is somehow subsumed by human culture and history or all culture and history is subordinate to and reflective of the character of the natural world.[73]

Hence the Environmental Humanities, in exploring both nature and culture—which, it turns out, are nearly the same thing—offer "a steady and persistent interrogation of the very meanings and definitions of the earth, of human artistic expression, and of humanity itself."

The idea is to magnify nature from mere subject of thought (e.g. the natural sciences) to the whole of thought itself. The division between what is human (and therefore has complex self-awareness, moral agency, a sense of beauty, and intimations of the transcendent) and

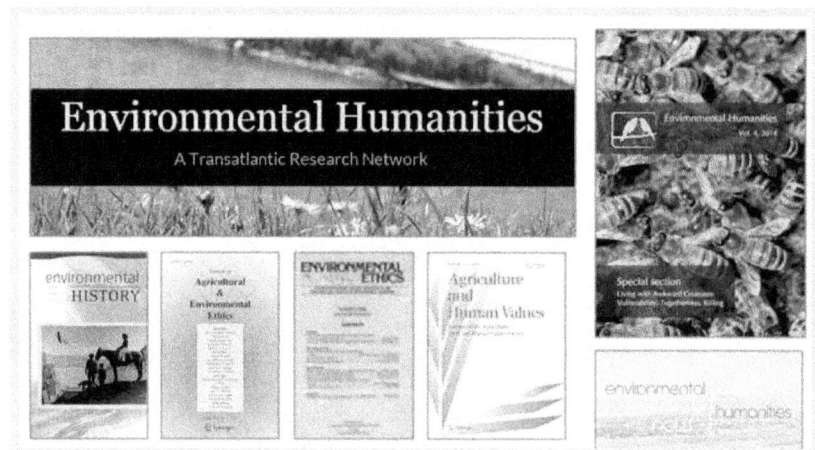

The relationship between human society and the natural world is the subject of many journals and societies.

what is outside the human in a "state of nature" is to be abolished, according to this view, and replaced with a conception that the "human" is just an eddy in the larger stream of existence.

To put this another way, natural scientists want to study man's influence on nature, and the more

---

72   George Handley, "What Are the Environmental Humanities?" Home Water, *Patheos*, October 26, 2012. http://www.patheos.com/blogs/homewaters/2012/10/what-are-the-environmental-humanities.html

73   *Ibid.*

orthodox environmental humanists want to study nature's influence on mankind. But this new branch of environmental humanism wants to promote the "Anthropocene," an epoch of history in which man and nature blur to the point of being indistinguishable. That's because man has so tampered with the environment, interrupting natural cycles and injecting pollution into the atmosphere, that when scientists examine nature, they actually are looking at extensions of our human existence. The role of the environmental humanities is to rejoin the arts and the sciences in order to take off the disciplinary blinders, take in a 360-degree view of the new human/nature reality, and, in a kind of undoing of the Socratic turn, reunite natural science with moral philosophy.

Living Laboratory

With everything from the classroom to the dormitory to the quad to the recycling center infused with opportunities to teach environmentalism, the campus quickly turns into a kind of training ground for sustainability. Weissman, the president of Dickinson College, describes campuses as functioning as "'living laboratories' of sustainability for the application of ideas, skills, and values developed in the classroom."[74] Students can learn about renewable energy in their science classes and help construct wind turbines on campus, or absorb from their English classes a habitual reverence for diversity that they can then express in their membership in the multicultural student club. Assuming, of course, that campuses already have sustainability policies and equipment that students can practice on, Weissman keenly notes the potential for classroom activities and homework assignments to shape students' lives:

> Classroom discussion of sustainability issues readily yields important implications for what we loosely call student "lifestyle." And residential practices similarly can be used as vehicles for reflection on, and study of, broader issues such as consumption and policy.[75]

Sustainability requires experiential learning. What better place to practice than on campus?

In this conception, the college campus is not a shelter from exterior distractions where students can focus their time on learning to understand reality and live appropriately in it. Instead, college becomes a miniature of the exterior world, a microcosm of the macrocosm, in which students are trained to operate in the "real world" once they get there. Whether there is still time for a leisurely appreciation of poetry or human history or philosophy is beside the point. The convenience, of course, is that the settings of the microcosm can be controlled, and the students can be taught to adopt a certain conception of "normal" that they then carry with them when they graduate.

The University of Wisconsin-Oshkosh has made this "living laboratory" idea into its unofficial motto. Its

---

74   Weissman, pg3.

75   *Ibid*

newest sustainability plan calls for the campus to become a testing ground of sustainability practices, so that students can "understand what sustainability is all about and how they can apply it in their own lives and in their own communities," as Professor Stephanie Spehar, a sustainability advocate, puts it.[76] The campus newspaper reported on the methods by which the university planned to make sustainability part of students' everyday lives:

> The plan calls for campus to become "a living learning laboratory" while infusing sustainability into the curriculum, developing a sustainability leadership program, creating incentives for student and faculty research and leveraging campus assets for the larger community.

The danger that the University of Wisconsin hopes to guard against is that without opportunities to learn about sustainability in class, students won't realize the environmental efforts taking place around them on campus, and without opportunities to practice on those campus initiatives, they might not ever come to apply the sustainability principle they learn in class.

To avoid this catch-22, the new 2014 Campus Sustainability Plan calls for more "co-curricular programming specifically focused on sustainability," and recommends incorporating sustainability into new-student orientation and other activities for first-year students, installing kiosks that highlight campus sustainability, and using sustainability efforts in campus advertising and promotional material.[77] It has already held recycling competitions, hired students to promote sustainable lifestyles to their peers, started an internship program at the sustainability office, and incorporated sustainability lessons into the "core concepts" taught in the general education program, the University Studies Program (USP). Beginning in Fall 2013, every student must take at least one course that answers the "Sustainability Signature Question," How do people understand and create a more sustainable world?, in order to be sure every student gets a taste of sustainability during his time in college.[78]

For these efforts, the University earned a 100 percent rating from the Association for the Advancement of Sustainability in Higher Education on the datum "Campus as a Living Laboratory" for its efforts to teach students about recycling, energy efficiency, diversity and inclusion, public transportation, and other ways to embody the sustainability virtues they were learning in class.[79]

---

76   Brook Wetor, "University Holds Discussion on new Sustainability Strategies," *Advance-Titan*, October 26, 2013. http://www.advancetitan.com/news/article_17ffc776-3e50-11e3-9e24-001a4bcf6878.html

77   "Co-Curricular and Residence Life," University of Wisconsin-Oshkosh Sustainability.

78   "Curriculum and Research," University of Wisconsin-Oshkosh Sustainability.

79   "University of Wisconsin-Oshkosh: AC-8: Campus as a Living Laboratory," AASHE STARS. https://stars.aashe.org/institutions/university-of-wisconsin-oshkosh-wi/report/2172/AC/curriculum/AC-8/

NAS

Enchanted

Students are not the only ones who change as a result of their sustainability educations. The professors change too. At Emory's Piedmont Project, Barlett soon found that when professors integrated sustainability into their courses, they changed as much, perhaps more, than their students did. Barlett, an anthropologist, uses the idea of "enchantment" to describe the transformation that takes place when, through the influence of sustainability, a person falls in love with nature. She tallies 184 faculty members who have participated in Piedmont Project workshops, in addition to 130 graduate students, for a total of 34 of 43 Emory departments that have at least one sustainability-influenced course. She believes that most, if not all, of these individuals and departments have changed to some degree as a result of their new experience with sustainability.

Peggy Barlett

For a 2008 article in *Cultural Anthropology*, Barlett reread all the email feedback surveys from previous Piedmont workshops and conducted one- to two-hour interviews with the first 37 participants from the Project's early years. Barlett concluded that a number of participants had experienced "reenchantment" with nature, because they had adopted stronger sustainability-related household and work habits as a result of their newly-discovered love for the earth.[80]

Barlett recounts some of the ways the professors changed as a result of their new teaching material: "For most faculty, the workshop stimulates curricular innovations and new personal actions, both at the household level and at work."[81] More specifically, participants noted that they had changed the way they thought about their lives and their jobs:

> The Piedmont Project workshop has probably been the most meaningful and deeply satisfying
> experience I have had...to shape my course...as well as reevaluate my role as an ducator.
> I realized we ought to work to make this place [the college] a sustainable way of living.
> It really did change the way I think.[82]

Seventy-three percent of the Project participants reported changing some action or habit of their daily lives (improving family recycling, controlling water run-off, choosing environmentally-friendly vacations), and 78 percent reported becoming more aware of sustainability and environmental issues at work (turning off computers, walking to work, even using departmental chair influence to encourage other colleagues

---

80   Barlett, Peggy F. "Reason and Reenchantment in Cultural Change: Sustainability in Higher Education," *Cultural Anthopology*, Volume 49, Number 6, 2008, pg. 1087.

81   Barlett, pg. 1077.

82   *Ibid.*

NAS

to change).[83]

Barlett attributes these changes to a "reenchantment" with nature sparked by a newly developed community working together towards sustainability:

> *Especially because the Piedmont Project did not seek directly to promote action but rather focused oncurriculumandpedagogy, whataccountsfor the power oftheshifts? In the participants' accounts and in the process of change over the past seven years, it is clear that the combination of reason and reenchantment is important. When participants talk about the experience, three things are emphasized: the supportive community that emerged, the new knowledge they enjoyed learning, and the new connections to place.[84]*

That sustainability metanarrative had done its work. The professors had changed the way they thought about some of their deepest values.

The Back Story

Higher education has only recently become the leading champion of sustainability. Twenty years ago, environmentalist David Orr worried that higher education might have the opposite effect: the average student, not deeply entrenched on either side of the issue, might become an environment trampler rather than a sustainability trumpeter.

David Orr

"Education is not widely regarded as a problem, although the lack of it is," Orr wrote in 1994. He was the Paul Sears Distinguished Professor of Environmental Studies and Politics at Oberlin College and a well-established figure among the academic environmental movement. His book, *Earth in Mind: Education, Environment, and the Human Prospect*, made the case for environmental education as an antidote to mainstream academic culture. "The conventional wisdom holds that all education is good, and the more of it one has, the better.... The truth is that without significant precautions, education can equip people merely to be more effective vandals of the Earth."[85]

Orr was writing in an intermediate lull after the heyday of the campus environmental movement of the 1970s and during the embryonic stages of the college campus sustainability movement. John Kerry and

---

83   Barlett, pg. 1082.

84   Barlett, pg. 1083.

85   David W. Orr. *Earth in Mind: Education, Environment, and the Human Prospect*, 10th Anniversary Edition, Washington: Island Press, 2004, pg. 5.

Teresa Heinz had met two years before at the 1992 UN Rio Summit on sustainability. Second Nature, their nonprofit founded to nurture the inchoate sustainability movement in American higher education, had just begun operation, and Chase and Rowland were just planting the seeds of the Ponderosa Project at Northern Arizona University. At the time, Orr saw academia as taking a lethargic, perhaps even wary, approach to sustainability. He wrote to expose what he perceived as a disordered focus on individual self-actualization at the expense of environmental action.

Fast forward two decades to 2014, when according to AASHE's calculations, American higher education institutions offered nearly 1,500 sustainability degree programs, hundreds more courses as electives, and thousands of co-curricular sustainable living programs. In a mere twenty years, sustainability has gone from a fringe concern to a central educational purpose.

What happened in between was one part coincidence and two parts strategy. The coincidence came by way of increasing public concern over global warming and burgeoning Western consumption. That concern was compounded by a series of extreme weather events. The 2004 Indian Ocean earthquake and tsunami, hurricanes Katrina in 2005 and Sandy in 2012, the BP oil spill in 2010, the meltdown of Japan's Fukushima Daiichi nuclear power plant in 2011 after the Tohoku earthquake and tsunami all turned the public mind towards the prevention of and reasons for environmental catastrophes. These harrowing experiences prepared the ground for the seeds of sustainability's solutions.

A media campaign, highlighted by a number of high-profile movies, interpreted these events through the lens of global warming and the solution of sustainability-minded action. The 2004 film *The Day After Tomorrow*, though fictitious, encapsulated American fears that global warming would melt the ice sheets, interrupting water flows, and leading to catastrophic global cooling. Al Gore's documentary *An Inconvenient Truth* (2006) rocked public awareness with charts, figures, and statistics meant to confirm the possibility of some of the doomsday scenarios that *The Day After Tomorrow* depicted. One year later in 2007, just as the Intergovernmental Panel on Climate Change was releasing its doleful Fourth Assessment Report, Leonardo DiCaprio tried his own hand at another documentary, *The 11th Hour*, that galvanized public interest in curbing global warming by downsizing hundreds of thousands of individual environmental footprints. "Global warming is not only the number one environmental challenge we face today, but one of the most important issues facing all of humanity," DiCaprio's voice warned the viewer. "We all have to do our part to raise awareness about global warming and the problems we as a people face in promoting a sustainable environmental future for our planet."

NAS

The combined effect was to capture the attentions and fears of the American populace and to cultivate increasing interest in finding a solution in the doctrines of sustainability. Meanwhile, two simultaneous sustainability strategies successfully capitalized on increasing public environmental interest.

The first strategy involved the United Nations, which announced the years 2005-2014 as its Decade of Education for Sustainable Development. This plan, largely carried out by UNESCO, sought to make lessons about global warming, disaster risk, poverty, diversity, gender equality, health, peace, water, and biodiversity central to K-12 education in countries around the world.[86]

The second was domestic, and it focused on higher education. Second Nature led the effort to develop a national strategy to get sustainability inside college syllabi, not just the college president's administrative agenda. This "Education for Sustainability" agenda grew out of a latent dissatisfaction with the way higher education had previously approached sustainability. In 2003, Second Nature co-founder and then-president Anthony Cortese expressed exasperation at higher education's unwillingness to engage with sustainability as an educational endeavor, despite a decade of work by Second Nature to convince them to do so. He wrote a short manifesto, "The Critical Role of Higher Education in Creating a Sustainable Future," in the journal *Planning for Higher Education*, in which he echoed many of Orr's fears from nine years earlier:

> *Despite the efforts of many individuals and groups within the formal educational system, education for a just and sustainable world is not a high priority. Indeed, it is the people coming out of the world's best colleges and universities that are leading us down the current unhealthy inequitable, and unsustainable path. Only a few architecture schools have made sustainable design a foundation of education and practice.*[87]

Three years later, Second Nature launched the American College and University Presidents' Climate Commitment with great success, as within a few years hundreds of institutions pledged to eliminate their greenhouse gases. But apart from the brief note about the benefits to colleges who are "integrating sustainability into their curriculum," the pledge did not actually commit signatory institutions to making any changes in their curricula.

Because of higher education's recalcitrance to make sustainability a part of their educational programs, students went on to graduation and to careers having never given much thought to their duties to the environment—

---

86   "Education for Sustainability," United Nations Education, Scientific, and Cultural Organization. http://www.unesco.org/new/en/education/themes/leading-the-international-agenda/education-for-sustainable-development/

87   Anthony Cortese. "The Critical Role of Higher Education in Creating a Sustainable Future," *Planning for Higher Education*, March-May 2003, pg. 16.

despite massive, multi-million dollar efforts on the part of their *alma maters* to cut out greenhouse gas emissions and to reduce water, paper, and energy usage. While hundreds of colleges and universities were making great strides towards becoming models of sustainability devotees, their tactics were akin to those any small corporation might take. The efforts focused on measurable administrative and operational goals, rather than student learning. They had achieved lower levels of energy and resource usage and purchased new emissions-reducing gadgets, but besides setting themselves up as examples, and advertising their achievements to prospective students, the institutions hadn't necessarily engaged their students in the process. The problem, as Orr saw it, was one of "green operations and brown curricula."[88]

Cortese called for an approach to sustainability in higher education that involved professors and students as well as presidents and executive staff. Only a college, with its "unique academic freedom and the critical mass and diversity of skills to develop new ideas, to comment on society and its challenges, and to engage in bold experimentation in sustainable living," could intervene fast enough to shift the mindset of a generation.[89] Cortese cites Orr: "The kind of education we need begins with the recognition that the crisis of global ecology is first and foremost a crisis of values, ideas, perspectives, and knowledge, which makes it a crisis of education, not one *in* education."[90]

Prepositions and Sustainability

Cortese set out a strategy that involved changing the content, context, and process of learning so that higher education began teaching "interdisciplinary systems thinking, dynamics, and analysis" in a context where environmental sustainability became "a central part of teaching of all the disciplines, rather than isolated as a special course or module in programs for specialists." This would be part of a process that emphasized "active, experiential, inquiry-based learning and real-world problem solving." Finally, Cortese held that none of these partial curricular changes would work unless higher education itself put on the practices of sustainability, and downsized its environmental footprint. "The university is a microcosm of the larger community," Cortese explained. "Therefore, the manner in which it carries out its dailyactivities is an important demonstration of ways to achieve environmentally responsible living and to reinforce desired values and behaviors in the whole community."[91]

Cortese's vision, despite the early setbacks with the Presidents' Climate Commitment, focusing on operational rather than educational goals, did eventually catch on. Bardaglio, Weissman, Barlett, Chase, Rowland, and the other early sustainability pioneers have brought the curriculum into the sustainability

---

88 Cited in Weissman, pg. 1.

89 Cortese, pg. 17.

90 Orr, 1994. Cited in Cortese, pg. 17.

91 Cortese, pg. 19.

line. Weissman has actually recommended a new version of the Presidents' Climate Commitment, one more focused on academics and student learning.[92]

These efforts were aided by the work of Stephen Sterling, the British professor of sustainability who categorized three levels of potential sustainability education, each one growing progressively more radical and closer to the ideal he hoped to achieve. Bardaglio cited Sterling's work as a helpful rubric for gauging progress in his "Moment of Grace" manifesto. First order sustainability education he called "education about sustainability," or the mere transmission of knowledge. Education about sustainability included science classes about sustainable species growth and resource usage, or seminars on the ways that pollution particles affect rain cycles. This, Bardaglio noted, "is easily assimilated within the status quo," as it requires merely adding more courses and programs to the academic roster, without tampering with any of the other courses and programs.[93] This first order of sustainability education has been largely accomplished.

Second order learning, or "education *for* sustainability," emphasized "learning for change," or educational tactics that got students actively practicing the things they were learning. Education for sustainability became the theme of a national convention hosted by Second Nature and the Campaign for Environmental Literacy in fall 2010. Twenty-three national sustainability leaders met in Washington, D.C. to lay out an *Education for Sustainability Blueprint*.[94] One subcommittee was assigned to the topics curriculum and research to figure out how to make higher education more interested in teaching students the habits and practices of sustainable living. The Blueprint commitment aimed to

1. Develop partnerships to develop and distribute curricular units.

2. Establish faculty development efforts; network of faculty leaders (e.g. through fellowships).

3. Evaluate and support the role of the president in leading this change, building on conversations that are already underway through the ACUPCC.

4. Develop connections with corporate America to show need for fundamental educational reform.[95]

Education for sustainability involved changing behaviors and operations to live and work in a sustainable

---

92   Weissman, pg. 6.

93   Bardaglio, pg. 17.

94   Education for Sustainability Blueprint, pg. 3

95   *Ibid*, pg. 7.

manner—the kinds of things that many universities, in signing the Presidents' Climate Commitment, have done. Most universities are well along the path towards this second order of sustainability education. The language of "living laboratories" and many of the tactics of psychological "nudging" fall into this category.

Third order learning focused on "education *as* sustainability." In this final phase, sustainability suffuses the entire pedagogical practice of the institution, so that sustainability operates as an assumption in every course, extracurricular activity, administrative policy, and lifestyle decision. It becomes a matter of practice, rather than a matter of study. It also embodies a type of pedagogy, so that, as Weissman explained, the manner of instruction becomes more active on the part of the student, more focused on group learning and research as ways to "create" knowledge, and more focused on synthesis and integration of disciplines rather than drawing distinctions among types of knowledge.

Here is where sustainability education battles are now waged. Education about and for sustainability are largely the norm on college campuses. Education as itself an exercise in sustainability, one whose pedagogy and inherent assumptions embody the principles of sustainability, is an idea still being sorted out. It has gained substantial ground, in the form of the Ponderosa and Piedmont projects and their progeny, and in places such as the University of Wisconsin-Oshkosh, where the whole campus aspires to embody the practices of sustainability. UW-Oshkosh, in its Campus Sustainability Plan that emphasized "living laboratory"-like initiatives, explains the motive behind sustainability as a pedagogic practice:

> *The clear links between sustainability and real-world problems encourages high-impact pedagogical practices suchasproblem-basedlearning, community and service learning, applied projects, and research. This in turn encourages us to transform ourthinking about learning at our institutions.*[96]

> *In this final phase, sustainability becomes a matter of practice, rather than a matter of study*

The lofty interdisciplinary goals of the Environmental Humanities carry the ideals of "education as sustainability" further still.

Harnessing higher education into the service of sustainability seriously undermines its purpose. It treats other disciplines as mere material for sustainability to interpret or vehicles by which sustainability can be taught. It forces habits and disciplines based on reflection, dialogue, and careful consideration into the

---

96   "Curriculum and Research," University of Wisconsin-Oshkosh. http://www.uwosh.edu/sustainability/csp-1/curriculum-and-research

mold of urgent political and social advocacy. It divorces the classroom from the goals of understanding and comprehending reality and yokes them to activism and ideological conformity. It cloaks the dogmas of environmentalism as necessary, foundational premises of higher education, setting them up as pillars that are above rational debate. And in refocusing the college curriculum on a popular politically-correct fad, it deprives students of a connection to a greater tradition of thought and culture. Eventually, though, sustainability will run out of liberal arts pillars to gnaw on and undermine. That is a habit, they will find, that cannot be sustained.

CHAPTER 2: THE ANATOMY OF CSM

Sustainability did not originate on the college quad, or even in faculty offices. Instead, as we have suggested, it took form in the UN Brundtland Report and seeped into America's consciousness slowly. The story of its ascent involves a couple, a summit, and a new idea.

In 1990, when John Kerry met Teresa Heinz at an Earth Day rally in Washington, D.C., sustainability was not part of the values commonly shared by the American public. Activists had celebrated the first Earth Day in April 1970 and prompted President Nixon to create the Environmental Protection Agency in December. These benchmarks were part of a larger environmental awakening that included the passage of the National Environmental Policy Act (1969), Clean Air Act (1970), Clean Water Act (1972), Marine Mammal Protection Act (1972), Endangered Species Act (1973), and the Safe Drinking Water Act (1974). President Nixon, more concerned with swing voters than with the merits of environmental protection, signed the bills. But under Reagan and Bush senior, environmental concerns played second fiddle to economic growth, international affairs, and tax policy. Sustainability was not even a well-known term.

Kerry, a Democratic senator from Massachusetts, spoke at the 1990 rally, as did Mrs. Heinz's husband, Republican Senator John Heinz from Pennsylvania. Senator Heinz introduced his colleague to his wife briefly, as the two waited to address the crowd. Senator Kerry and Mrs. Heinz met a second time in 1992, this time in Rio de Janeiro at a UN Earth Summit called to address escalating fears of climate change, the fourth in a series of UN climate talks around the world. Kerry was still a Massachusetts senator, divorced in 1988. Heinz, widowed in 1991 after her husband was killed in a helicopter crash, attended as a delegate from the U.S. State Department.

At the Earth Summit, delegates adopted a Climate Change Convention that led to the emissions-cutting Kyoto Protocol in 1997. They also developed *Agenda 21*, an action plan for nations, local governments, and private organizations to strengthen "sustainable development" by combating poverty, empowering women and minorities, providing food and resources for children, protecting biodiversity, conserving virgin land, and setting out educational goals to teach citizens about the importance of sustainable programs of action. The U.S. technically adopted *Agenda 21* when President George H.W. Bush signed it at the Summit, but the document was not legally binding and did not require formal ratification by the Senate. President Bush declined to enforce it. If sustainability were to become prevalent in American thinking, private advocacy groups would need to succor it.

One such advocacy group was the International Council for Local Environmental Initiatives (ICLEI). ICLEI-

NAS

Local Governments for Sustainability smuggled the principles of *Agenda 21* into unwitting townships and counties. But that tactic provoked outrage in Oklahoma, Iowa, Missouri, Maine, Tennessee, New Hampshire, Alabama, Arizona, and Kansas, all of which took action to prevent the implementation of the *Agenda* in their states.

Kerry and Heinz left the Summit united by a different idea. They decided to target colleges and universities as seedbeds of the idea, lending sustainability the sheen of professional scholarly validation and training a new generation in its values and ambitions. Higher education already provided a haven for left-leaning ideas outside the mainstream, and a growing number of ecologists, scientists, philosophers, and poets were beginning to organize into a loose network of environmentally-conscious intellectuals. While other organizations, such as ICLEI, attempted to slip *Agenda 21* into local governments and city planning meetings, higher education could afford to be more forthright in its goals. One year later Kerry and Heinz launched a nonprofit, Second Nature, with the mission to "create a sustainable society by transforming higher education."[97]

Kerry and Heinz married in 1994, and their brainchild, Second Nature, went on to become a major force in creating the American sustainability movement. Second Nature started by targeting professors willing to introduce sustainability matter into their courses (adding units and themes here and there in non-environmental courses). Second Nature also encouraged the creation of new centers of sustainability study. But the most powerful ally turned out to be college presidents themselves. In 2006, at a joint meeting with the heads of the Association for the Advancement of Sustainability in Higher Education (AASHE) and the advocacy group ecoAmerica, Second Nature commissioned one of the most powerful agents of the American sustainability movement: The American College and University Presidents' Climate Commitment.

The idea was simple. College presidents possessed the power to set their institutions' agendas and

---

97   "Mission," Second Nature. http://www.secondnature.org/mission

NAS

to introduce sustainability as a key principle. They had an unparalleled ability to shepherd the new movement to adulthood. And they had the financial flexibility to experiment with new technologies and programs. In December 2007, twelve presidents became founding signatories of the Commitment.[98] Some of the institutions in this first cohort were not prominent, but the Commitment snagged a handful of very important presidents and institutions, including Michael Crow, the president of Arizona State University, and Bernard Machen, the president of the University of Florida.

The group of twelve pledged that they could "recognize the scientific consensus that global warming is real and is largely being caused by humans" and affirmed the need to slash global greenhouse gas emissions by 80 percent by 2050 (at latest) in order to "avert the worst impacts of global warming and to reestablish the more stable climatic conditions" that they believed characterized previous history.[99] They avowed that American energy independence was needed "as soon as possible." And they testified that higher education was uniquely responsible to convince the broader public of these facts and to set an example by going carbon-neutral. "We believe colleges and universities must exercise leadership in their communities and throughout society by modeling ways to minimize global warming emissions, and by providing the knowledge and the educated graduates to achieve climate neutrality," the preamble read. It continued,

> *Campuses that address the climate challenge by reducing global warming emissions and by integrating sustainability into their curriculum will better serve their students and meet their social mandate to help create a thriving, ethical and civil society.*[100]

They pledged to audit all campus-caused greenhouse gas emissions, identify ways to cut back, and develop a target date and strategy to eliminate or offset 100 percent of all campus emissions. They also promised to complete at least two of seven activities to draw attention to sustainability: mandate that all new campus buildings meet LEED silver certification for efficiency; purchase only ENERGY STAR- certified efficient appliances; offset (by purchase of carbon credits) all emissions caused by institutional air travel; encourage or provide public transportation for all students, staff, and faculty; buy energy from renewable sources; engage in shareholder activism to pressure the corporations in which the college owned stock to move towards climate neutrality; and host a recycling competition among the students.

---

98 "Mission and History," American College and University Presidents' Climate Commitment. http://www. presidentsclimatecommitment.org/about/mission-history. The twelve founding presidents and their institutions are: Loren Anderson, President, Pacific Lutheran University; Michael Crow, President, Arizona State University; Nancy Dye, President, Oberlin College; Jo Ann Gora, President, Ball State University; David Hales, President, College of the Atlantic; Bernard Machen, President, University of Florida; Gifford Pinchot III, President, Bainbridge Graduate Institute; Kathleen Schatzberg, President, Cape Cod Community College; Mary Spilde, President, Lane Community College; Douglas Treadway, President, Ohlone College; Darroch Young, Chancellor, Los Angeles Community College District; Paul Zingg, President, California State University, Chico.

99   "Text of the American College & University Presidents' Climate Commitment," American College and University Presidents' Climate Commitment. http://www.presidentsclimatecommitment.org/about/commitment

100   "Text of the American College & University Presidents' Climate Commitment."

NAS

The pledge was a success. By March 2007, another 140 had signed on as charter signatories.[101] As of this writing (January 2015) 685 colleges and universities have taken the pledge to eliminate all carbon-based emissions and to train their students in the habits of sustainability.

Launching the CSM

Second Nature conjured a movement out of emotional passion and smart strategies. Sustainability is now among the highest priorities at colleges and universities even in an age beset by financial pressures. Colleges and universities teach sustainability in their departments (1,438 sustainability-focused academic programs at 475 campuses[102]) and

Figure 2. Map of AASHE STARS Members

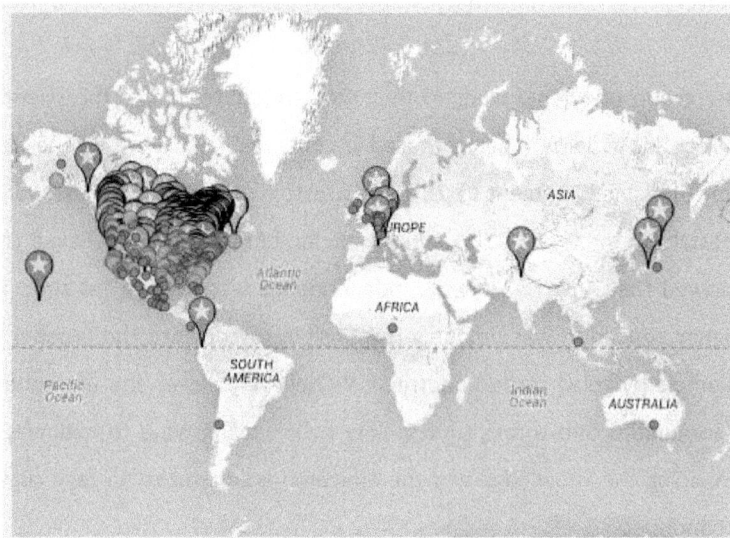

incorporate principles of sustainability into non-sustainability themed courses. They center residence life programs and student life initiatives around sustainability. They hire campus directors and staff—nearly 500, according to a 2012 AASHE survey— tasked with making the campus more sustainable and making the students better fitted to sustainability's mold.[103] They spend hundreds of millions of dollars cutting down on carbon emissions, trash, and electricity and water usage. Even colleges and universities that have not formally signed the President's Climate Commitment have followed the trend to make sustainability central to their operations.

Every year, 322 universities vie for lead places in *The Princeton Review*'s "Guide to Green Colleges," a report first added to the line-up of *Princeton Review* guides in 2007.[104] They pledge institutional affiliations by membership in the Association for the Advancement of Sustainability in Higher Education (772 institutions worldwide, 694 in the United States),[105] or by joining the Association of University Leaders

---

101 "Mission and History," American College and University Presidents' Climate Commitment.

102 "AASHE Academic Programs Database," Association for the Advancement of Sustainability in Higher Education. http://www.aashe.org/resources/academic-programs/

103 "Salaries and Status of Sustainability Staff in Higher Education," AASHE, 2012, pg. 2. http://www.aashe.org/files/documents/programs/2012_staffsurvey-final.pdf

104 *The Princeton Review*'s Guide to Green Colleges. http://www.princetonreview.com/green.aspx

105 "AASHE Member Directory," Association for the Advancement of Sustainability in Higher Education. http://www.aashe.org/membership/member-directory?keyword=&field_organization_type_value_many_to_one%5B%5D=Two+Year+Institution&fie

for a Sustainable Future (400)—not to mention signing the Presidents' Climate Commitment.[106] Colleges and universities (664 institutions, 587 of them in the US) fastidiously track their progress on various sustainability goals by reporting to AASHE's STARS program,[107] which awards bronze, silver, and gold stars for meeting progressive levels of sustainability achievements. STARS awards points for actions such as incorporating sustainability into new student orientation (2 points),[108] growing an organic garden (0.25 points),[109] offering professors incentives to complete research on sustainability (6 points),[110] stocking the dining hall with recycled paper napkins (0.25 points),[111] pursuing clean or renewable energy (7 points),[112] subsidizing child care for employees (0.25 points),[113] and offering gender neutral housing (0.25 points).[114]

Universities audit their greenhouse gas emissions and regularly report to the Presidents' Climate Commitment on their progress towards eliminating emissions. They assess their students on international Sustainability Literacy Tests and compare their performances to students at other colleges.[115] They ask new students to take a Sustainability Pledge upon matriculation and to repeat it again at graduation. The Pledge at the University of Virginia, for instance, reads in full,

> *I pledge to consider the social, economic, and environmental impacts of my habits and to explore ways to live more sustainably during my time here at U.Va. and beyond.*[116]

Emory University goes further and asks students to take at least three action steps from a prescribed list, including turning off unused computers and disabling screensavers, choosing stairs over elevators, studying only in well-populated places at night to conserve energy for lighting, and taking "some time for

Id_organization_type_value_many_to_one%5B%5D=Four+Year+Institution&field_organization_type_value_many_to_one%5B%5D=Graduate+Institution&country%5B%5D=us&province

106   Association of University Leaders for a Sustainable Future. http://www.ulsf.org/

107   "STARS Participants & Reports," Association for the Advancement of Sustainability in Higher Education. https://stars.aashe.org/institutions/participants-and-reports/?sort=country

108   "ER-3: Sustainability in New Student Orientation," Association for the Advancement of Sustainability in Higher Education. https://stars.aashe.org/institutions/auburn-university-al/report/2013-01-15/ER/co-curricular-education/ER-3/

109   "ER-T2-2: Organic Garden," Association for the Advancement of Sustainability in Higher Education. https://stars.aashe.org/institutions/auburn-university-al/report/2013-01-15/ER/co-curricular-education/ER-T2-2/

110   "ER-18: Sustainability Research Incentives," Association for the Advancement of Sustainability in Higher Education. https://stars.aashe.org/institutions/auburn-university-al/report/2013-01-15/ER/research/ER-18/

111   "OP-T2-10: Recycled Content Napkins," Association for the Advancement of Sustainability in Higher Education. https://stars.aashe.org/institutions/auburn-university-al/report/2013-01-15/OP/dining-services/OP-T2-10/

112   "OP-8: Clean and Renewable Energy," Association for the Advancement of Sustainability in Higher Education. https://stars.aashe.org/institutions/auburn-university-al/report/2013-01-15/OP/energy/OP-8/

113   "PAE-T2-4: Childcare," Association for the Advancement of Sustainability in Higher Education. https://stars.aashe.org/institutions/auburn-university-al/report/2013-01-15/PAE/human-resources/PAE-T2-4/

114   "PAE-T2-1: Gender Neutral Housing," Association for the Advancement of Sustainability in Higher Education. https://stars.aashe.org/institutions/auburn-university-al/report/2013-01-15/PAE/diversity-and-affordability/PAE-T2-1/

115   The Sustainability Literacy Test. http://www.sulite.org/en/sustainability_home

116   "Pledge with Us," Sustainability at the University of Virginia. http://www.virginia.edu/sustainability/take-the-pledge-2/

stillness once a week."[117]

In addition to times of contemplation and self-examination, there are also times of celebration. A number of sustainability holidays have cropped up on campus calendars and student schedules. Campus Sustainability Day has been celebrated every year since 2003, when the Society for College and University Planning organized the day as a tribute to new developments in the sustainability movement. Now run by AASHE and Second Nature, Campus Sustainability Day is observed on the fourth Wednesday in October and centers around a yearly theme and webcast keynote address. (The theme for 2014 was "Empowering Change on Campus and in the Community" and for 2013 was "Climate Adaptation: Resilient Campuses and Communities.")

Seventy schools participated in Campus Sustainability Day 2014.[118] James Madison University held a public trash sort to rescue recyclables.[119] Students at Arizona State University watched *Carbon Nation*, a film about the potential of transitioning away from a carbon-based economy, observed a solar sewing machine demonstration, and participated in the energy saving competition "Green Team – Kill Your Vampire Energy."[120] At the University of Iowa, student environmental groups set up stations with surveys for their peers to calculate their carbon footprint. Newly aware of their ecological weight, students could snap a photo with a whiteboard, where they'd written goals aimed at shrinking their carbon use.[121]

America Recycles Day (November 15) is a national event sponsored by the nonprofit Keep America Beautiful to encourage Americans to pledge to increase their recycling rates. Many colleges and universities hold their own events to participate, often using an America Recycles Day guide to encouraging recycling at tailgating parties.[122] During its 2014 celebration, Georgia State University held a recycling fest on the quad with music, free food, a workshop on reusing t-shirts, and a bottle toss competition.[123] Auburn University brought out Aubie the mascot in a green shirt emblazoned with the recycling symbol. Auburn students who signed a recycling pledge were entered into a raffle to win a football autographed by the

---

117   "Make Your Personal Sustainability Pledge," Sustainability Initiatives, Emory University. http://sustainability.emory.edu/cgi-bin/MySQLdb?VIEW=/viewfiles/view_pledge.txt&pageid=1042

118   "This Year's Events," Campus Sustainability Day, October 22, 2014. http://campussustainabilityday.org/this-years-events-2/

119   "James Madison University," Campus Sustainability Day, October 22, 2014. http://campussustainabilityday.org/2014/10/22/james-madison-university/

120   "Arizona State University," Campus Sustainability Day, October 22, 2014. http://campussustainabilityday.org/2014/10/22/arizona-state-university/

121   "University of Iowa: Carbon Foot Print Photo Campaign," Campus Sustainability Day, October 20 – October 24, 2014. http://campussustainabilityday.org/2014/10/22/university-of-iowa/

122   "Go-to Guide: Tailgating Recycling Events," America Recycles Day, 2013. http://americarecyclesday.org/wp-content/uploads/2012/06/Tailgating-Go-to-Guide-20132.pdf

123   "Georgia State University-America Recycles Day," America Recycles Day, Keep America Beautiful, November 14, 2014. http://events.americarecyclesday.org/event/georgia-state-university-america-recycles-day/

football coach.[124]

There is also Global Divestment Day on February 13-14, sponsored by 350.org to pressure institutions to divest from fossil fuels and to encourage students to keep protesting. During Dark Sky Week in April, activists–including some at universities—shut off building lights at night to raise awareness about light pollution. They also want to decrease fossil fuel energy use so that from space, their geographic region will look like developing regions around the world: dark, and electricity-starved. The irony, as American

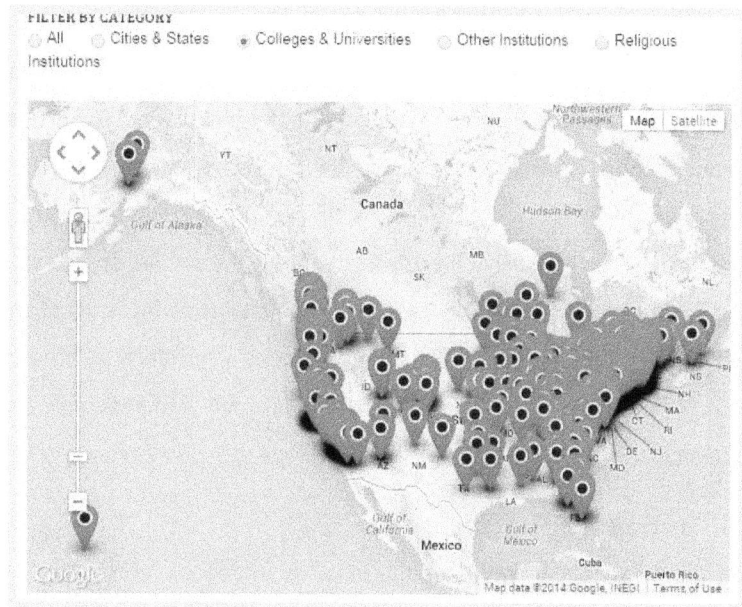

Figure 3. Map of Campuses with Fossil Free Divestment Campaigns

University professor Caleb Rossiter pointed out in a *Wall Street Journal* op-ed, is that developing nations want to look like America, brightly lit up with electricity and the advantages it brings.[125]

Many students have taken sustainability to heart. They work as interns and "eco-reps" for their university's Office of Sustainability,[126] volunteer to dig through their dorm trash bins to track recycling rates, monitor their peers' food waste,[127] write senior theses on strategies to reach carbon neutrality,[128] march on campus demanding more stringent environmental policies,[129] and, in the case of Harvard, blockade the president's office if she does not move fast enough to reshape the college's endowment according to principles

---

124   "Auburn University America Recycles Day," America Recycles Day, Keep America Beautiful, November 14, 2014. http://events.americarecyclesday.org/event/auburn-university-america-recycles-day/2015-11-14/

125   Caleb Rossiter, "Sacrificing Africa for Climate Change," *Wall Street Journal*, May 4, 2014. http://www.wsj.com/articles/SB10001424052702303380004579521791400395288

126   "Eco-Rep Program," University of Massachusetts-Amherst. http://www.umass.edu/sustainability/get-involved/eco-reps

127   "Waste Audit Allows Students to Get Their Hands Dirty," *Today at Colorado State*, February 28, 2014. http://today-archive.colostate.edu/story.aspx?id=9738

128   Carol Ness, "Sustainability Award for Senior Thesis on Campus Emissions," *UC Berkeley News Center*, October 22, 2012. https://newscenter.berkeley.edu/2012/10/22/sustainability-award-for-senior-thesis-in-campus-emissions/

129   Joy Resmovits, "Fossil Fuel Divestment Campaign Escalates at Swarthmore," *Huffington Post*, May 17, 2013. http://www.huffingtonpost.com/2013/05/17/fossil-fuel-swarthmore_n_3294687.html

of environmental and social sustainability.[130] Roused by the charismatic environmentalist Bill McKibben, students at more than 400 universities have started "Go Fossil Free" campaigns asking their alma maters to divest from coal, oil, and natural gas. Sustainability has succeeded pacifism, diversity, feminism, and economic justice as the newest campus radical movement.

Of course, many colleges and universities that have not signed the American College and University Presidents' Climate Commitment nonetheless are strong supporters of the movement and have taken many of the same steps. Harvard president Drew Faust, for instance, has declined to take the Commitment, yet nevertheless has led the university to expand sustainability-related course offerings, reward faculty for their research on sustainability, slash greenhouse gas emissions, establish the university's first Sustainability Committee, and a hire a vice president for sustainable investing.[131] She, and many others, hesitate to sign up not out of conviction that the Commitment is mistaken in either its diagnosis or its prescriptions, but out of financial prudence. To sign the Commitment is to announce an institution's determination to move ahead with some difficult and expensive steps to "eliminate all greenhouse gas emissions."

Dissecting the Movement

At times, sustainability can appear as a spontaneous grassroots movement: People get tired of smog and worried about rising temperatures, and they start developing new tastes for "greener" products. The market responds by providing cleaner manufacturing technology, credential programs for sustainability professionals, and green brand names to meet consumer demand.

On one level, that's true. Market forces will naturally react and cater to the tastes and preferences fostered by new social movements. But the sustainability movement is more than a coincidental shift in millions of Americans' economic calculus. Behind that shift is an organizational structure that introduces sustainability into educational curricula, stages political battles, and sponsors a public awareness campaign.

In addition to the handful of organizations already mentioned, scores of nonprofits, for-profits, professional societies, and journals and magazines tag-team to keep the sustainability message prominent in people's minds. There are philanthropic organizations focused on K-12 education (Center for Ecoliteracy, Center for Environmental Education) and higher education (Second Nature, Association for the Advancement of Sustainability in Higher Education). Some cater to students (Energy Action Coalition, California Student Sustainability Coalition), others directly to the general public (Green America). There are organizations for women (Women's Network for a Sustainable Future), "indigenous" peoples (Indigenous Environmental

130 Amna H. Hasmni, "Undergraduate Protester Arrested for Blocking Entrance to Mass. Hall," *Harvard Crimson*, May 2, 2014. http://www.thecrimson.com/article/2014/5/1/divest-protester-arrested-mass-hall/

131 Drew Faust, "Confronting Climate Change," Letter to the Harvard Community, Office of the President, Harvard University, April 7, 2014. http://www.harvard.edu/president/news/2014/confronting-climate-change

NAS

Network), and religious groups (Earth Quaker Action Team).

Some focus directly on particular battles. As You Sow, Responsible Endowments Coalition, Divestment Student Network, and 350.org's Go Fossil Free work on fossil fuel divestment campaigns. The NYC Environmental Justice Alliance and Green for All promote racial diversity within the environmental movement. The U.S. Green Building Council awards certificates for buildings that meet sustainable construction standards.

There are clubs for kids, such as Earth Force and Earth Guardians. And there is a host of professional associations for adults: the Association of Sustainability Practitioners, the Sustainability Management Association, the International Society of Sustainability Professionals, Sustainable Food Trade Association, Sustainable Living Association, Association of Christian Sustainability Professionals, and the Sustainability Consortium. And in case our society fails in its efforts to become "sustainable" and is forced to "adapt" to this brave new world of climate change, there's even the American Society of Adaptation Professionals (conveniently, the urgent backronym ASAP) prepared to step in and set us back on a sustainable track.

Some professional associations are focused specifically on higher education. In addition to joining the Presidents' Climate Commitment and AASHE, colleges and universities can become members of the Alliance for Resilient Campuses (also sponsored by Second Nature), the Higher Education Associations Sustainability Consortium, and International Sustainable Campus Network. Administrators can join the Association of University Leaders for a Sustainable Future. Professors can join the International Society for Environmental Ethics, the American Society for Environmental History, or the International Society for the Study of Religion, Nature, and Culture. They can also attend academic conferences such as the annual Sustainability Ethics and Entrepreneurship conference at the University of Denver.

Increasingly scholars have diverted their attention to the topic of sustainability as historians and sociologists have begun to chronicle sustainability's rise from obscurity a quarter century ago to its recent popularity. Historian Jeremy Caradonna from the University of Alberta offers *Sustainability: A History* from Oxford University Press (September 2014), tracing the idea of "sustainability" to the German 18th century *nachhaltigkeit*. On the other hand, *Sustainability: A Cultural History* by the German Ulrich Grober and translated by Ray Cunningham (UIT Cambridge, 2012) attempts to date sustainability to ancient Greek philosophy. *The Principles of Sustainability* by Simon Dresner (Routledge, 2008) offers prescriptions for how to get an unsustainable society to some plateau of sustainability. In a 2012 article for the *Hedgehog Review*, University of Virginia sociologist Joshua J. Yates argues that sustainability has emerged as a "master term" for our time, a word that "provides a perspicuous lens through which to view the ethical

NAS

disposition and emotional temper of a culture at a particular moment in time." [132]

There are national and regional meetings where sustainability devotees can swap tales and advise their peers. For ten years, the University of Maryland has hosted a Smart and Sustainable Campuses Conference for sustainability directors at colleges and universities. AASHE holds an annual national conference billed as the "largest gathering of higher education sustainability professionals and students in North America" with approximately 2,000 attendees in 2014. [133] AASHE also sponsors local consortia by state and by region. [134] Second Nature holds a Presidential Summit on Climate Change each year to stir presidents' sustainability impulses.

Sustainability has its own journal, *Sustainability: The Journal of Record*, launched in 2008 in partnership with AASHE. Other journals include *International Journal of Sustainability in Higher Education*; *Journal of Sustainability Education*; *International Journal of Sustainable Built Environment*; *Sustainability: Science, Practice, & Policy*; and *International Journal of Sustainable Development*.

Meanwhile, for the laypeople, there are green magazines and news sites, such as Triple Pundit, GreenBiz, and Eco Watch. *Insider Higher Ed* sponsors a "Getting to Green" blog offering serialized "sustainability commandments" to sustainability-minded administrators. And for the general public, major news publications give sustainability a friendly welcome. *The New York Times* runs the online "Dot Earth" blog as well as friendly stories in the news print edition. *The Guardian* has launched a "Sustainable Business" blog. *Bloomberg News* has a sustainability editor, Eric Roston; so do *Architects Journal* and *Seafood News*.

Businesses have begun taking up the sustainability mantle as well. Stores ranging from Target[135] and Walmart[136] to upscale boutiques[137] have introduced "sustainable" product lines. A survey from *GreenBiz* found that the number of firms with executive sustainability officers jumped from 67 in 2005 to 283 in

---

132 Joshua J. Yates, "Abundance on Trial: The Cultural Significance of 'Sustainability,'" *The Hedgehog Review*, Vol. 14 No. 2, Summer 2012. http://www.iasc-culture.org/THR/THR_article_2012_Summer_Yates.php

133 "Empowering Higher Education Sustainability Professionals," AASHE Conference and Expo. http://conference.aashe.org/2014/

134 "State and Regional Campus Sustainability Organizations," AASHE. http://www.aashe.org/resources/state-and-regional-campus-sustainability-organizations

135 "Sustainable Products," Target Corporate Responsibility. https://corporate.target.com/corporate-responsibility/environment/sustainable-products

136 "Walmart's Great Value Brand Goes Green with New Cleaning Products," Press Release, Walmart, November 8, 2013. http://news.walmart.com/news-archive/2013/11/08/walmarts-great-value-brand-goes-green-with-new-cleaning-products

137 Modavanti Sustainable Fashion Boutique. https://modavanti.com/sustainability

2012.[138] A host of sustainability consultants, ranging from Sustainalytics, American Sustainable Business Council, Sustainable Business Consulting, Strategic Sustainability Consultant, and dozens more have popped up to market their sustainability expertise.

Even organizations that traditionally have been more focused on the environment alone rather than on sustainability have begun to come around. Greenpeace, with a budget of more than $33 million[139] and with 250,000 members in the United States (2.8 million members worldwide),[140] explains in its "About Us" web page that it aims to "create solutions that promote environmental sustainability rooted in social justice."[141] Greenpeace's USA Executive Director, Annie Leonard, is one of the most familiar faces in the sustainability movement. Leonard created the "Story of Stuff" and "Story of Bottled Water" videos that have gone viral as campus sustainability directors use them to teach students to recycle their waste, buy fewer things, and quit drinking bottled water. Meanwhile, the 1.4 million-member-strong Natural Resources Defense Council, started in 1970 to promote and help draft environmental legislation, begins its mission statement, "We seek to establish sustainability and good stewardship of the Earth as central ethical imperatives of human society."[142] Even the National Wildlife Federation, which works to protect endangered species and habitats, has phrased its work in the language of sustainability: among its three goals for preserving wildlife is "Forming resilient and sustainable solutions to problems facing our environment and wildlife."[143]

In addition to grassroots social movements, the sustainability crowd has been active in political battles. The Sierra Club and 350.org have fought the Keystone XL pipeline, organizing sit-ins, protests, and mass-emailing and -calling campaigns. There have also been efforts to sway voters. During the 2014 midterm elections, environmental sustainability groups spent upwards of $85 million on campaigns. California billionaire Tom Steyer alone spent $72 million through his PAC, NextGen Climate Action Committee.[144] In what Gene Karpinski, president of the League of Conservation Voters, called "by far the biggest investment that the environmental community has ever made in politics,"[145] the LCV footed the

---

138   John Davies, "State of the Profession," *GreenBiz*, January 2013, pg. 6. http://ugs.utah.edu/sustainability-certificate/State%20 of%20the%20%20Profession%202013.pdf

139   Greenpeace, 2013 990 tax form. http://www.greenpeace.org/usa/Global/usa/planet3/PDFs/GP-FY13-Public%20 Disclosure%20Copy%20Form%20990.pdf

140   "Our Work," Greenpeace. http://www.greenpeace.org/usa/en/about/

141   "Our Commitment to Diversity and Inclusion," Greenpeace. http://www.greenpeace.org/usa/en/about/

142   "Mission," Natural Resources Defense Council. http://www.nrdc.org/about/mission.asp

143   "Who We Are," National Wildlife Federation. http://www.nwf.org/Who-We-Are.aspx

144   NextGen Climate Action, 2014 PAC Summary Data, *Open Secrets*, as of November 24, 2014. http://www.opensecrets.org/pacs/lookup2.php?strID=C00547349

145   Chris Mooney, "Environmental Groups Are Spending an Unprecedented $85 Million in the 2014 Elections," *Washington Post*, October 27, 2014. http://www.washingtonpost.com/blogs/wonkblog/wp/2014/10/27/environmental-groups-are-spending- an-unprecedented-85-million-in-the-2014-elections/

NAS

bill for a slew of smear ads attacking candidates who "deny" climate change.[146] There are also the Sierra Club PAC, with about $1 million,[147] and the Environmental Defense Action Fund, at about $3 million.[148]

Polls indicate that these efforts are trying to transform, rather than cater to, the interests of society. A September 2014 Pew Research Poll showed the environment as eighth among eleven issues most important to voters.[149] In November 2014, voters soundly rejected all but two of six Senate candidates and one of five gubernatorial candidates backed by environmentalist groups. Nor are consumers naturally that much greener than voters. In fact, over the last few years "eco," "green," and "environmentally-friendly" have begun triggering negative responses in consumers, and boutique suppliers are rebranding as luxury or health companies. In August, the *Wall Street Journal* reported that small businesses especially were downplaying any environmentalist connections in order to build a reliable customer   base.[150]

But among youth, the movement has made its deepest inroads. A March 2014 Gallup poll found that 43 percent of all adults "worried" about global warming "only a little/not at all." But 38 percent of those ages 18-29, and 41 percent of those ages 30-49 worried about it "a great deal."[151] Likewise, a March/April 2014 survey from Harstad Strategic Research found that 69 percent of Millennials thought the government should be "more involved" in "addressing the issue of climate change or global warming." An overwhelming 80 percent supported "a national policy requiring electric utilities to generate at least one-third of their power from renewable energy sources like wind and solar by 2030," and 79 percent favored policies to "reduce carbon pollution to deal with climate change and global warming." Fifty percent ranked "climate & renewables" as one of the top issues that would determine how they voted.[152]

The generation gap that has opened up on the issue of sustainability is almost entirely the result of the successful efforts of advocates to plant the movement in schools and colleges.

---

146   Videos, League of Conservation Voters. http://www.lcv.org/media/video/

147   Sierra Club, 2014 PAC Summary Data, *Open Secrets*, November 24, 2014. https://www.opensecrets.org/pacs/lookup2.php?strID=C00135368

148   Environmental Defense Action Fund, *Open Secrets*. http://www.opensecrets.org/outsidespending/detail.php?cmte=C90014895&cycle=2014

149   "Wide Partisan Differences Over the Issues That Matter in 2014," *Pew Research Center for the People and the Press*, September 12, 2014. http://www.people-press.org/2014/09/12/wide-partisan-differences-over-the-issues-that-matter-in-2014/

150   Amy Westervelt, "Small Firms Are Downplaying Their Green Side," *Wall Street Journal*, August 25, 2014. http://www.wsj.com/articles/small-firms-are-downplaying-their-green-side-1408912053

151   Frank Newport, Americans Show Low Levels of Concern on Global Warming," Gallup, April 4, 2014. http://www.gallup.com/poll/168236/americans-show-low-levels-concern-global-warming.aspx

152   "National Online Survey of Millennial Adults – March/April 2014," Harstad Strategic Research Inc., March-April 2014.

NAS

Diversifying the Movement

Caring about the environment has often been a mark of relative wealth; only the prosperous have money to shell out for more expensive organic food and new gadgets such as solar panels or high efficiency refrigerators. And only those with stable, comfortable lives have time to devote to raising public awareness about the fate of the natural environment. It's hard to care about the ozone layer when you're cashing out a payday loan to cover the rent.

Hundreds of thousands of demonstrators marched for "climate justice" at the People's Climate March on September 21, 2014, in New York City, two days before the UN held a climate summit there.

Recently, though, as the environmental movement has morphed into the sustainability movement, the cause has taken on new social and economic dimensions. The sustainability movement has actively recruited the poor and minorities to join the campaign, broadening its message into a catch-all for environmental *and* social grievances. Economic inequality, racial discrimination, sexism, and egalitarian democracy have all become issues at home within sustainability.

The history behind the shift is telling (we will get into more detail later) but it's important to at least note these changing demographics within the movement. The largest public environmentalist demonstration to date, the People's Climate March in September 2014, billed itself as an "environmental justice" campaign to address "environmental racism." More than 400,000 people in 162 countries took to the streets in advance of a UN meeting in New York City (leading up to the UN climate change conference in Paris, 2015).[153] The main march, down the west side of Manhattan, was organized by Eddie Bautista of the NYC Environmental Justice Alliance, which according to its mission seeks to link "low-income neighborhoods and communities of color" to protest against the disproportionate effects of pollution, climate change,

---

153   People's Climate March. http://peoplesclimate.org/

NAS

and damage from natural disasters that these communities face.[154]   The march line-up was broken into six segments representing different demographics; first in line came "indigenous, environmental justice, & other frontline communities" to represent those first affected by climate change.

Figure 4. The Lineup for the People's Climate March in New York City[155]

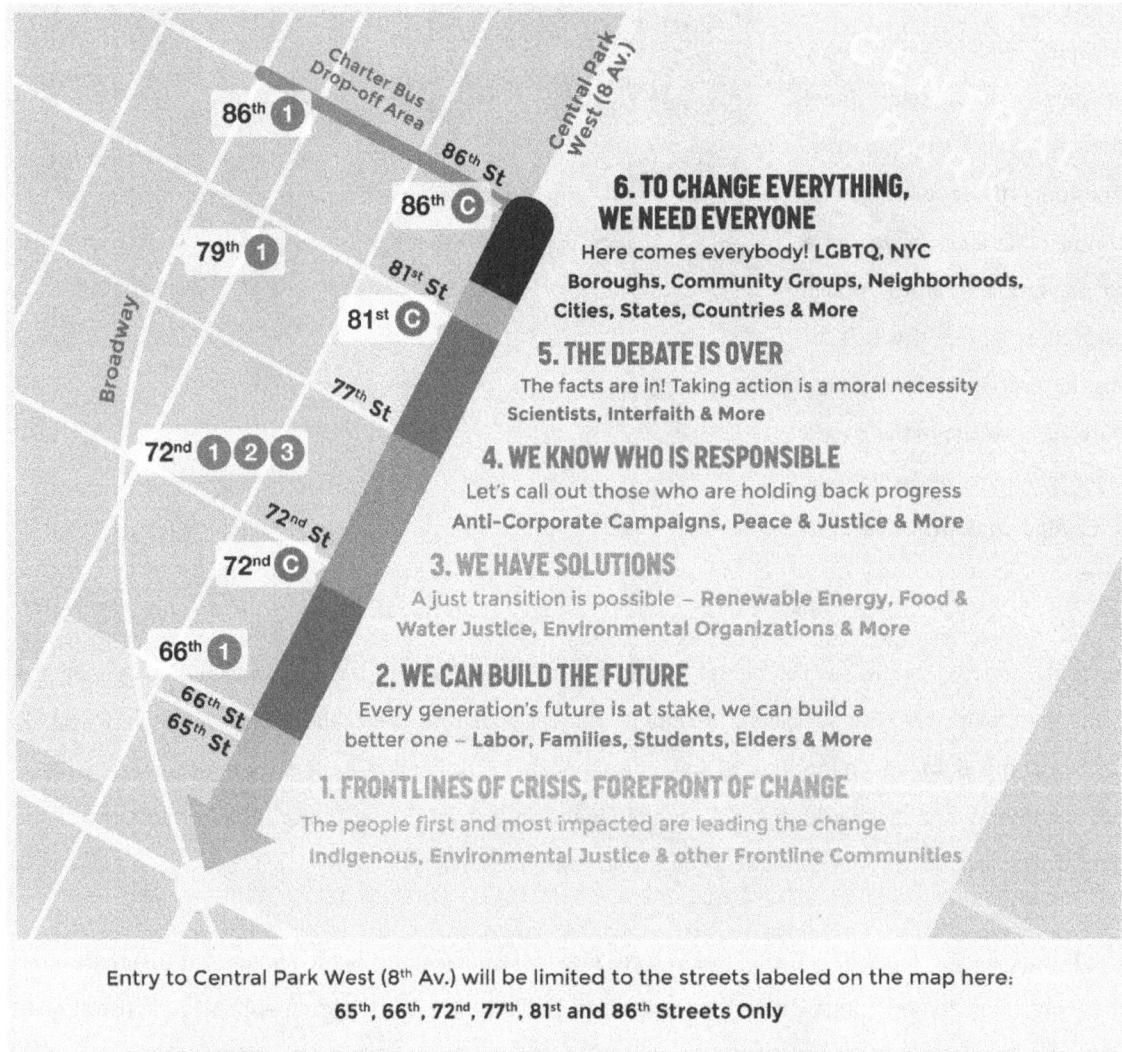

Entry to Central Park West (8th Av.) will be limited to the streets labeled on the map here:
65th, 66th, 72nd, 77th, 81st and 86th Streets Only

The Climate March was not the first time the minority and sustainability movements converged. When racial turmoil swept the nation in summer 2014 following the deaths of two black men—Michael Brown in Ferguson, Missouri, and Eric Garner in Staten Island, New York—at the hands of white police officers, the sustainability movement reached out to angry minorities. Deirdre Smith from 350.org wrote a blog

---

154   "Mission," New York City Environmental Justice Alliance. http://nyc-eja.org/?page_id=25

155   "The People's Climate March Lineup," The People's Climate March. http://peoplesclimate.org/lineup/

post, "Why the Climate Movement Must Stand with Ferguson," that was widely quoted and reprinted by other environmental and sustainability groups. In the piece, Smith, a black woman, wrote, "To me, the connection between militarized state violence, racism, and climate change was common-sense and intuitive."[156] The common factor linking all three was the prejudice of state actors and corporations whose racism led them to mistreat minorities, "dehumanize" them in low-wage jobs, locate landfills and dirty factories in poor, often minority neighborhoods, and force the poor and unprepared to deal with "disproportionate" effects of climate change and natural disasters.

Because all the battles were linked to capitalism and racism, fighting separate battles for a higher minimum wage, immigrant amnesty, police reform, pollution limits, and greenhouse gas emissions caps would never work. The individual movements needed to combine. "The fossil fuel industry would love to see us siloed into believing that we can each win by ourselves on 'single issues,'" Smith wrote. But those single-issue divisions "prevent us from building the movement we need to create a new future for ourselves, a future where we have clean energy that doesn't kill us, and creates jobs that provide dignity and a living." She called for the environmental movement to "dig deep" and get serious about promoting more minorities to movement leadership and to join the protests in Ferguson: "Part of that work involves climate organizers acknowledging and understanding that our fight is not simply with the carbon in the sky, but with the powers on the ground."[157]

In August 2014, the Sierra Club established an award named after Robert Bullard, father of the environmental justice movement. Bullard, remarking on the award, called for activists to make 2015 the "year of diversity" within the environmental movement.[158] Then in December, Sierra Club executive director Michael Brune wrote in a *Huffington Post* article that after he began speaking against police racism and for President Obama's executive decision to not deport 5 million illegal residents, some Sierra Club members chided him for getting involved with these "non-environmental" issues. "Here are some quick thoughts on why these issues are so important to address," Brune responded:

> *Injustices in our political system and in our culture empower polluters and lead to the destruction of our most cherished places. Those same injustices often breed hatred, sow division among us, and threaten our health and safety. The Sierra Club's mission is to "enlist humanity" to explore, enjoy, and protect the planet. That mission, which applies to everyone, cannot be achieved when people's rights are being violated and their safety and dignity are being threatened on a routine basis. This*

---

156   Deirdre Smith, "Why the Climate Movement Must Stand with Ferguson," *350.org*, August 20, 2014. http://350.org/how-racial-justice-is-integral-to-confronting-climate-crisis/

157   Ibid.

158   Brentin Mock, "How Environmental Justice Fared in 2014 — and the Outlook for 2015," *Grist*, December 31, 2014. http://grist.org/politics/how-environmental-justice-fared-in-2014-and-the-outlook-for-2015/

must stop. [159]

A week later Naomi Klein took Brune's logic one step further in a *Nation* article, "Why #BlackLivesMatter Should Transform the Climate Debate." Referencing the Twitter hashtag that activists had used while protesting the deaths of Michael Brown and Eric Garner, Klein argued that racism not only affected the environment (and thus was a natural target of the environmental movement), but that racism had become one of the central enemies of environmental progress. "What does #BlackLivesMatter, and the unshakable moral principle that it represents, have to do with climate change?" Klein asked. Her answer? "Everything." [160]

Klein's article recounts Hurricane Katrina, when predominantly black Americans endured days without food and potable water but the rest of the country failed to curb carbon emissions in an effort to prevent more extreme weather events, and Typhoon Haiyan, which devastated the Philippines but hasn't stopped Canadians from pulling oil out of their tar sands. At the UN climate convention at Copenhagen in 2009, nations agreed to limit warming to 2 degrees Celsius—a temperature Klein deemed dangerous to the Earth, and fatal to some island and African nations. Delegates from the presumed future victim nations began chanting "2 degrees is suicide" and "We will not die quietly." Western nations better situated and prepared for a warming planet had discounted the lives of people in other nations, Klein charged.

"Taken together, the picture is clear," Klein summed up. Denial of climate change, resistance to environmental regulations, and reticence to engage in climate treaties signified racism:

> *Thinly veiled notions of racial superiority have informed every aspect of the non-response to climate change so far. Racism is what has made it possible to systematically look away from the climate threat for more than two decades. It is also what has allowed the worst health impacts of digging up, processing and burning fossil fuels—from cancer clusters to asthma—to be systematically dumped on indigenous communities and on the neighborhoods where people of colour live, work and play…. If we refuse to speak frankly about the intersection of race and climate change, we can be sure that racism will continue to inform how the governments of industrialized countries respond to this existential crisis. It will manifest in the continued refusal to provide serious climate financing to poor countries so they can protect themselves from heavy weather. It will manifest in the fortressing of wealthy continents as they attempt to lock out the growing numbers of people whose homes will become unlivable.* [161]

---

159   Michael Brune, "Why the Sierra Club Can't Be Silent," *Huffington Post*, December 6, 2014. http://www.huffingtonpost.com/michael-brune/why-the-sierra-club-cant_b_6282530.html

160   Naomi Klein, "Why #BlackLivesMatter Should Transform the Climate Debate," *The Nation*, December 12, 2014. http://www.thenation.com/article/192801/what-does-blacklivesmatter-have-do-climate-change#

161   Ibid.

A number of activists in the sustainability movement have taken the climate change-racism message to heart. Ian Monroe, CEO of Oroeco, a company that helps people track their sustainability footprint, wrote in the *Huffington Post*, "Climate change is not primarily an environmental issue....Climate change is the greatest racial and social justice issue of our time" (bold in original).[162] The Sustainability Studies Office at the University of Minnesota, noting the "heavy intersection of the issues of race relations and sustainability," has compiled a series of resources on the Ferguson protests for students to read.[163] The Responsible Endowments Coalition, which has been helping to lead the fossil fuel divestment movement, published a statement of "Solidarity with Ferguson and All People Resisting Systemic Racial Violence."[164] The California Student Sustainability Coalition released a similar statement that maintained, "our struggle and liberation is indelibly bound to the liberation of others. We cannot have climate justice or a sustainable planet without racial justice."[165]

These attempts to blend tenets of sustainability and diversity signify a recent shift in the sustainability movement's strategy. The vexed Sierra Club members whom Michael Brune rebuffed in his *Huffington Post* article represent an older generation of environmentalists from a time when the environmental movement focused more exclusively on matters of the natural world. Sustainability, as a second wave of environmental thought and activism, is freer to intermingle.

Even so, the overlap between the sustainability and diversity movements is not seamless. On a theoretical level, sustainability and diversity are in many ways rival dogmas competing for the same conceptual space. Both claim to identify the primary problem plaguing society, invite single-minded devotion to the cause, and assume a moralistic posture.

Even proponents of environmental justice can have a hard time describing the conceptual relationship between the social and environmental aspects. When two Bowdoin College students organized a February 2014 conference called "Environmental Justice," a newswriter working on a post for the college website asked them to define environmental justice. One of the organizers, Courtney Payne, a junior, replied, "It's a little hard to define." The closest she could come to a definition was, "One thing that sticks out to me as part of a definition is that the people who are doing the most environmentally harmful things are not

162   Ian Monroe, "Fostering a Climate for Racial Justice," *Huffington Post*, December 30, 2014. http://www.huffingtonpost.com/ian-monroe/fostering-a-climate-for-r_b_6397440.html

163   Nathan Michielson, "Resources Regarding #Ferguson," Office of Sustainability Studies, University of Minnesota, August 20, 2014.   http://www.susteducation.umn.edu/2014/08/20/resources-regarding-ferguson/

164   Nina Macapinlac, "Responsible Endowments Coalition Stands in Solidarity with Ferguson and All People Resisting Systemic Racial Violence," *Responsible Endowments Coalition*, October 13, 2014. http://www.endowmentethics.org/responsible_endowments_coalition_stands_in_solidarity_with_ferguson_and_all_people_resisting_systemic_racist_violence

165   Zen Trenholm, "CSSC In Solidarity with Ferguson," California Student Sustainability Coalition, November 25, 2014. http://www.sustainabilitycoalition.org/cssc-in-solidarity-with-ferguson/

NAS

the ones who are seeing the biggest consequences a lot of the time."[166]

The rivalry between racial and environmental concerns plays out both in campus culture and in academic discourse, where the relationship is rather one-sided, with sustainability eager for an ally and diversity wary of being preyed upon. Students at Swarthmore College told us that Mountain Justice, the student environmentalist club agitating for fossil fuel divestment, was eager to become an ally and in some cases spokesman for the minority- and diversity-themed groups, but that the relationship was "tense." 350.org and the Sierra Club may be eager to paint racial diversity as a sub-issue of environmentalism and sustainability, but the advocates of affirmative action and immigration reform do not, for the most part, endorse the description. One of us noted some years ago the competitive relationship between the diversity and sustainability ideologies in a *Chronicle of Higher Education* article, "From Diversity to Sustainability: How Campus Ideology Is Born."[167]

But for those like Brune and Klein, who do accept the tangled triumvirate of social, environmental, and economic concerns that characterize sustainability, diversity's tie-in is clear. Racial discrimination is seen as an enabler for industrial pollution by perpetuating the low wages that make cheap production viable. Environmental degradation is perceived as entrenching racial injustice by condemning poor minority communities to blighted, barren lands.

Gender and Sustainability

From linking climate change and environmental irresponsibility with racism to blaming it for economic inequality, homophobia, and patriarchy is only a short mental jump. Some scholars are already drawing the connections between sustainability and gender issues. Bailey Kier, a Ph.D. candidate in American Studies at the University of Maryland, College Park, is completing a dissertation on "An American River: A Queer Geography of the Potomac River Basin and Environmentalism in the Nation's Capitol." A transgender man, Kier speaks and writes on topics such as "Imaginaries of Estrogenic Ecocatastrophe: Transsexuality and Systemic Hormonal Ecologies in an Anthroheterocentric World,"[168] and "Queering Agriculture."[169] Agriculture, in Kier's mind, is a field ripe for "queer" analysis, because modern agriculture centers on modifying the genetic material and reproduction of crops. He positions himself outside the mainstream sustainability movement, but prods the movement to recognize its ideological affinity with

166 Rebecca Goldfine, "Two Students Bring Environmental Justice Debate to Bowdoin," Bowdoin College, January 27, 2014. http://community.bowdoin.edu/news/2014/01/two-students-bring-environmental-justice-debate-to-bowdoin/

167 Peter Wood, "From Diversity to Sustainability: How Campus Ideology Is Born," *Chronicle of Higher Education*, October 3, 2010. http://chronicle.com/article/From-Diversity-to/124773/

168 Bailey Kier, "Imaginaries of Estrogenic Ecocatastrophe: Transsexuality and Systemic Hormonal Ecologies in an Anthroheterocentric World," Gender, Bodies, and Technology Conference, Virginia Tech, May 1-3, 2014. http://www.cpe.vt.edu/gbt/gbt2014program.pdf

169 Bailey Kier, "Queering Agriculture: Food Security in the Nation's Capital and the Crises of Reproductive American Familism," University of California-Berkeley, Center for the Study of Sexual Culture, February 10, 2015. http://cssc.berkeley.edu/events/event/queering-agriculture-food-security-in-the-nations-capital-and-the-crises-of-reproductive-american-familism/

NAS

progressive gender studies. Kier is offended that the movement discusses organic, local food in the context of traditional family farms. "Family" has sexist, heterosexual connotations. Or, as Kier puts it, "the growing popularity of sustainable food is laden with anthroheterocentric assumptions of the 'good life' coupled with idealized images and ideas of the American farm, and gender, radicalized and normative standards of health, family, and nation."[170]

Kier acknowledges that he is on the edgier side of the sustainability movement. But even activists in the mainstream of the movement view gay rights, gender parity, and the dissolution of the traditional family as kindred issues with the sustainability movement. "Save the Earth, don't give birth"[171] is one slogan of the movement. Another, "wrap with care, save the polar bear," is emblazoned on condoms that one environmental group distributes for free.[172]

Women's issues are forefront in the minds of many sustainability activists and well-established in foundational documents in the sustainability movement. Diane Macheachern, an environmental researcher and activist, calculates that elevating women to political leadership could dramatically make societies more sustainable. Writing in "Women, Consumption, and Sustainable Development," a chapter in the anthology *Powerful Synergies: Gender Equality, Economic Development, and Environmental Sustainability* published by the United Nations Development Programme, Macheachern notes that after rising consumption of resources, gender inequality is the most powerful factor restraining sustainability. She explains,

> *A second—yet equally important—factor undermining sustainable development goals is directly linked to ongoing and pervasive gender inequalities.... This explicit acknowledgement that sustainable development goals cannot be actualized without both transforming consumption patterns and achieving gender equity underlies the emerging philosophy behind the green economy. "A green economy and any institutions devised for it must make their core focus the well-being of people—of all people, everywhere—across present and future generations," concludes a task force convened by the Pardee Center for the Study of the Longer-Range Future at Boston University in the U.S. "That essential idea puts the notion of equity... smack at the centre of the green economy enterprise."*[173]

Macheachern notes that deconstruction of the patriarchy may tangibly help the fight against climate change:

> *In general, women are more inclined than men to favour sustainability as a lifestyle choice*

---

170 Ibid.

171 Mark Bauerlein, "Save the Earth, Don't Give Birth," *First Things*, January 26, 2015. http://www.firstthings.com/blogs/firstthoughts/2015/01/save-the-earth-dont-give-birth

172 Tim Blair, "Save the Earth, Don't Give Birth," *Telegraph*, February 23, 2010. http://blogs.news.com.au/dailytelegraph/timblair/index.php/dailytelegraph/comments/save_the_earth_dont_give_birth/

173 Diane Macheachern, "Women, Consumption, and Sustainable Development," *Powerful Synergies: Gender Equality, Economic Development, and Environmental Sustainability*, United Nations Development Programme, September 2012, pp. 69-70. http://www.undp.org/content/dam/undp/library/gender/Gender%20and%20Environment/Powerful-Synergies.pdf

NAS

(GenderCC 2012). Research shows that this is true in poor and rich countries and regions alike.... Although their choices are influenced by income levels, social conditions and other biases, women are also motivated by their reproductive role and the impact their purchases could have on their families' long-term well-being (Johnsson-Latham 2007). Where men are more likely to turn to technological solutions, women demonstrate a greater willingness to change lifestyle behaviours and to consider the 'precautionary principle' in their day-to-day choices.[174]

Nor is the focus on gender issues new. The Earth Charter, drafted in 1994 and endorsed by more than 6,000 international organizations, including nearly 500 universities from around the world, affirms "social and economic justice" as the third of its four main points.[175] Signatories affirm their commitment to aims such as

- Eradicate poverty as an ethical, social, and environmental imperative.

- Affirm gender equality and equity as prerequisites to sustainable development.

- Eliminate discrimination in all its forms, such as that based on race, color, sex, sexual orientation, language, and national, ethnic, or social origin.

- Integrate into formal education and life-long learning the knowledge, values, and skills needed for a sustainable way of life.[176]

The United Nations, too, has embraced this flexible conception of sustainability as encompassing issues as broad as race and sexual orientation. In September 2015, the UN will announce its Sustainable Development Goals that will replace the 2000-2014 Millennium Development Goals. The working list of seventeen "sustainable development" goals includes environmental goals such as "Ensure availability and sustainable management of water and sanitation for all," "Ensure sustainable consumption and production patterns," "Take urgent action to combat climate change and its impacts," alongside such social and economic targets as "End poverty in all its forms everywhere," "Achieve gender equality and empower all women and girls," "Reduce inequality within and among countries," and "Promote peaceful and inclusive societies for sustainable development, provide access to justice for all and build effective, accountable and inclusive institutions at all levels."[177]

174   *Ibid*, pg. 71.

175   Earth Charter signatories, Earth Charter International, as of December 11, 2014. http://earthcharterinaction.org/database/endorsed.php

176   "The Earth Charter," Earth Charter International, June 29, 2000. http://www.earthcharterinaction.org/content/pages/Read-the-Charter.html

177   "Sustainable Development Goals," Open Working Group Proposal for Sustainable Development Goals, United Nations. https://sustainabledevelopment.un.org/sdgsproposal

NAS

A high profile coalition, "action/2015" of more than a thousand organizations has begun a campaign to encourage global leaders to agree to the UN's goals at the UN Special Summit on Sustainable Development in September 2015 and the UN Climate Talks in December. An open letter to "world leaders" (cc'd to "everyone else") asks them to "come up with a grand new global contract for our one human family" that represents the voices of every demographic: "the voice of a young girl currently deprived an education... of a pregnant mother deprived healthcare... of young people deprived decent work... of a family from a minority group fearful of discrimination from corrupt officials... of farmers forced to migrate to cities as climate refugees."[178] Action/2015 is headed by dozens of influential names such as Bill and Melinda Gates, Ben Affleck, Malala Yousafzai, Bono, Mo Ibrahim, Queen Rania Al Abdullah of Jordan, Columbia's Jeffrey Sachs, and, of course, one of the seminal founders of the idea of sustainability itself, Gro Harlem Brundtland.

The Brundtland Report

The central document that set the sustainability movement in motion and that united concern for the environment with a collection of other leftist ideologies was the 1987 UN report, *Our Common Future*, better known as the Brundtland Report. It remains a touchstone for sustainability advocates.

Gro Harlem Brundtland, three-time prime minister of Norway (1981; 1986–89; 1990–96), presented the finished report to the World Commission on Environment and Development. The document put forth a definition of "sustainable development" that has been cited ever since: "development that meets the needs of the present without compromising the ability of future generations to meet their own needs."

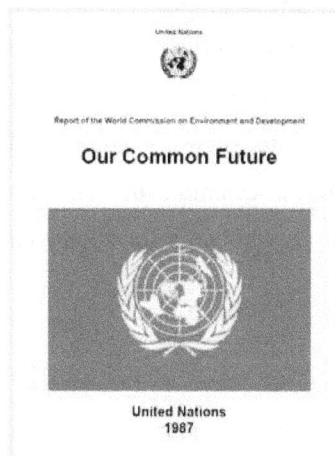

United Nations

Report of the World Commission on Environment and Development

Our Common Future

United Nations
1987

The definition gave sustainability its characteristic generational concern for the health of the future as much as for the health of the present. The definition is not the only one in circulation among sustainability advocates but it is far and away the most frequently cited.[179]

The Brundtland definition of sustainability shifted the orientation of development projects from the short-term to the long-term future. The word "development"—buttressed in the public vocabulary by terms such as "progress," "growth," and "advancement"—had for many years represented an ambitious, optimistic aspiration to make the world better and wealthier. The addition of the modifier *sustainable*

---

178   "Open Letter to World Leaders," *action/2015*. http://www.action2015.org/what-were-doing/world-leaders-letter/

179   For an analysis of several definitions of sustainability see Daniel Bonevac, "Is Sustainability Sustainable?" *Academic Questions*, Vol. 23, No. 1, Spring 2010, pp. 84-101.

NAS

announced a major change. "Sustainable development" aimed lower: its aspiration was simply to secure what is necessary to maintain daily life. And instead of aiming at increases in production to accomplish this, it aimed to cut consumption. The new, shorter horizon was to reduce the standard of living to a level that could be maintained on a plateau. In some cases the aim sounded desperate: averting a coming wholesale destruction precipitated by our present overuse of resources. Sustainable development saw the earth as an easily depleted endowment, rather than an investment fund. Rather than aiming to engineer better solutions to increase the total available wealth of the world and thereby increasing the wellbeing of everyone, sustainability called for using only renewable resources and husbanding them with extreme caution.

At the time of the Brundtland Report, the term "sustainable development" was an unhappy compromise, a lexical attempt to forge a parchment treaty between the interests of post-colonized countries eager for development, industrialization, and economic wellbeing, and wealthier Western countries worried about global warming.[180] Environmentalists initially hated the term for smuggling in language redolent of industrialism, and for suggesting that "development" provided a legitimate option. The environmental movement up to that point had called for a zero-growth, stable environment, in which population growth was dangerous (Paul Ehrlich's 1968 *The Population Bomb*), increased consumption of resources was risky (the Club of Rome's 1970 *Limits to Growth*), and development of any kind risked putting the earth past some tipping point. They saw "sustainable development" as a contradiction in terms, an unfortunate concession that threatened to water down the environmentalist fundamentals with free market—or at least consumerist—thought.

But in another sense, the birth of sustainability was closely linked to that older environmentalist heritage. The term may have been an uncomfortable fit for many in the environmental movement when Brundtland forced it upon them in 1987, but the word soon lost its industrial overtones, and within a few years, the word "development" disappeared from the concept entirely, at least in the United States. In Europe and the Commonwealth countries, the term "sustainable development" remains in common use. In the United States, it has become a term used almost exclusively by professionals who work specifically on development projects.

The Brundtland Report thus became known for its popularization and definition of the idea of "sustainability" alone, pure and simple. This new version of sustainability became known in the technical journals as "strong sustainability," in contrast to the Brundtland Report's original "weak sustainability."

---

180   Frances Aldson, "EU Law and Sustainability in Focus: Will the Lisbon Treaty Lead to 'the Sustainable Development of Europe'?" *School of Oriental and African Studies, London*, 2011, 23 ELM 5 284. http://www.lawtext.com/lawtextweb/default.jsp?PageID=2&PublicationID=6&pubSection=4#16

NAS

"Strong sustainability" discarded Brundtland's limited confidence that technology could expand the frontier of production possibilities by new methods and efficiencies. Instead it demanded that social and economic systems shrink to accommodate what were presented as narrow and ever narrowing natural constraints. By 1993, when Kerry and Heinz were laying the foundations for Second Nature, they could comfortably couch their mission in the language of advancing "sustainability" and trust that the environmental community would perfectly understand.

In that way, sustainability holds close ties to the environmental waves of the 1960s and 1970s. It could revere April 22, 1970, the first Earth Day, when 20 million Americans demonstrated across the country for environmental regulation. It could reference Rachel Carson's 1962 bestseller *Silent Spring*, in which Carson, a marine biologist turned popular science writer, argued that DDT and other pesticides were poisoning wildlife and might make springtime song birds extinct. And it could express equal indignation at oil spills near the coast of Great Britain in 1967 and Santa Barbara, California, in 1969, or the mercury poisoning of fishing waters near the Japanese

Rachel Carson

factory town of Minamata from 1932 to 1968. When in 1988 NASA scientist James Hansen testified before the U.S. Congress that global warming was happening, sustainability could claim this environmental catastrophe as one of its own issues to tackle.

Similarly, sustainability could handily reference early American Transcendentalism, when Ralph Waldo Emerson and Henry David Thoreau waxed eloquent about the solitary life of peace with nature in the 1830s, and the American conservation ethic in the latter half of the 1800s, with the founding of the Audubon Society in 1886 and the Sierra Club in 1892, and the first celebration of Bird Day in 1894. Teddy Roosevelt made the idea mainstream with his creation of national parks, and sustainability could retroactively baptize Roosevelt's preservationist impulse as an early version of its own desire to preserve havens of nature and ward off technological interference.

But in taking on the role of second-wave environmentalism, sustainability also differed from its predecessor in key ways. One of those ways was the distinct spin that sustainability gave environmental concerns. In the 1960s and 1970s, activists worried about unwanted additions to the environment: toxins in the stream, pesticides on food, trash on the sidewalks, and smog in the air. Sustainability shifted focus to unwanted extractions from the environment. No longer was the concern one of whether the stream water was clean or not, but whether it was there at all. Sustainability worried about over-taxing natural resources and expending them too quickly, leaving the Earth unable to replenish its local stores.

NAS

Perhaps the biggest difference, though, between environmentalism and sustainability was that sustainability self-consciously concerned itself with worldwide systems of social interaction and community, not just pollution. The Brundtland Report made this distinction clear when it linked the economy and issues of social equality to environmental preservation. "Development involves a progressive transformation of economy and society," the Report announced, in the paragraphs explaining its "meet the needs of the present" definition of sustainability.[181] Sustainability required a "concern for social equity between generations, a concern that must logically be extended to equity within each generation," making the immediate equality of distribution of goods and privileges a key part of the sustainability teaching.

Figure 5. The Sustainability Venn Diagram

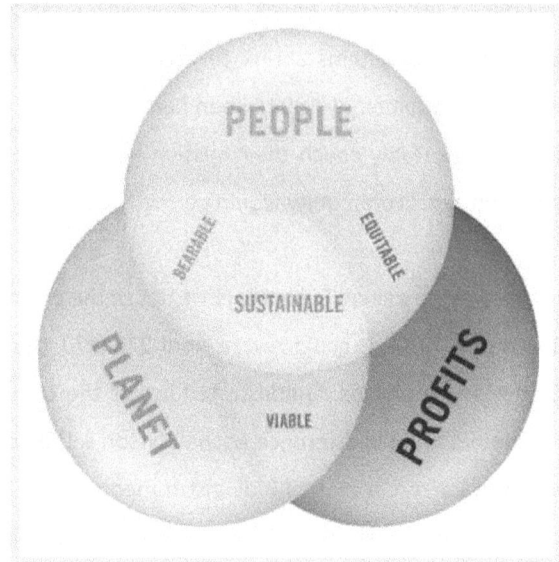

The University of South Florida-Saint Petersburg depecits sustainability in a popular Venn diagram.

Soon sustainability advocates laid out a tripartite rubric for understanding sustainability under the heads of Economy, Environment, and Society. This could be easily depicted on a Venn diagram, with the three categories as interlocking circles. The intersection of environmental and economic circles of concerns might represent the conservationist philosophies of the early 1900s. The merging of economics and society might indicate a moral economy with fairly distributed resources. And the overlap of social and environmental interests would denote a land ethic in which people felt connected to the earth and carried out their responsibility to protect it in exchange for the earth's provisions. But only "sustainability," in the center of the diagram, overlapped with all three. As Second Nature describes the issue,

> Too often, we view major societal issues (health, social, economic, security, and environmental) as separate, competing, and hierarchical, when they are really systemic and interdependent. We do not have environmental problems per se. We have environmental consequences resulting from the way we have designed our business, social, economic, and political systems.[182]

The new rubric of sustainability opened up the environmental movement to embrace the widest imaginable assortment of social and political concerns. Thus a sustainability event could become an all-purpose

---

181   *Our Common Future*.

182   "Mission," Second Nature. http://www.secondnature.org/mission

collection of grievances whose aims might include, for example, eradicating poverty and hunger, encouraging universal primary education, empowering women, freeing Tibet from Chinese rule, reducing child mortality, improving maternal health, and combating HIV/AIDS and malaria, as well as championing environmental care.

Demonstrators at the People's Climate March celebrated the earth's birthday by proposing a series of changes in environmental policy, early education, gender equality, healthcare, population control, and other social issues.

Social Ecology

Sustainability's social and economic interests do not spring from thin air of course, and like sustainability's environmental strand, they too have historical predecessors. Consternation about population equilibrium dates back centuries, most notoriously to the British economist Thomas Malthus (1766-1834). Malthus reasoned that because population increased geometrically (that is, exponentially), and agricultural yields increased at a merely arithmetic rate, population would inevitably outstrip the food supply, until famine and disease, or food wars and political turmoil, reduced the population to a manageable size.

Paul Ehrlich, a Stanford entomologist who wrote the 1968 best-seller *The Population Bomb*, helped return Malthusian thought to public esteem. Ehrlich suggested strict birth control and a severe reduction of foreign aid to poor countries. "The penalty for frantic attempts to feed burgeoning populations in the next decade may be a lowering of the carrying capacity of the entire planet to a level far below that of 1968," Ehrlich wrote coldly. "We are rapidly destroying our planet as a habitat for Homo Sapiens."[183]

Four years after Ehrlich's bestseller was released, a group of businessmen and academics known as the Club of Rome released a report from MIT scientists that foretold similar doom. *The Limits to Growth* explained computer model simulations that projected overpopulation, resource depletion, dwindling food sources, and increased pollution, unless society shifted to a "global equilibrium" at one-fourth the 1970 global population and the 1970 per-capita resource consumption. Bill McKibben wrote in *Eaarth* that that "few events in environmental history were more significant than the publication of that slim book, which was translated into thirty languages and sold 30 million copies, more than any other volume on

---

183   Cited in Rubin pg. 82.

the environment."[184] McKibben says his own environmental awakening came from reading *The Limits to Growth*.

But demographic data have disproven Malthusian hypotheses. Wealthy (and thus well-fed) countries tend to have fewer children, and poorer (hungrier) countries have more. Worldwide life expectancy has risen alongside worldwide population, thanks in great measure to the "green revolution" that improved farming techniques and bred heartier crops that have fed a worldwide population now approaching eight billion. Today "hunger" more often means poor nutritional habits rather than literal starvation.

History's rebuttal of Malthus aside, sustainability's social and economic concerns find heroes in other sources, too. There is Murray Bookchin (1921-2006), a former Stalinist turned anarchist, who sounded the toxin-alarm bell in his 1962 *Our Synthetic Environment* even as Rachel Carson popularized the fear in *Silent Spring*. Bookchin's interests, though, centered more on social and economic revolution than on environmental clean-up. In a 1964 essay, "Ecology and Revolutionary Thought," published under the pseudonym Lewis Herber, he introduced the term "social ecology" to indicate a healthy communitarian society that shared political and economic power and enjoyed environmental purity.[185] He elaborated in an essay, "What is Social Ecology?":

Murray Bookchin

> *What literally defines social ecology as 'social' is its recognition of the often overlooked fact that nearly all our present ecological problems arise from deep-seated social problems. Conversely, present ecological problems cannot be clearly understood, much less resolved, without resolutely dealing with problems within society. To make this point more concrete: Economic, ethnic, cultural, and gender conflicts, among many others, lie at the core of the most serious dislocations we face today—apart, to be sure, from those produced by natural catastrophes.*[186]

For Bookchin, the goal of environmentalism was not to save the planet per se but to create ultra-egalitarian human societies.

Barry Commoner (1917-2012), a professor of biology and a third party candidate in 1980 for President

---

184   Bill McKibben, *Eaarth: Making a Life on a Tough New Planet*, St. Martin's Griffin, 2011, pg. 90.

185   Lewis Herber (Murray Bookchin), "Ecology and Revolutionary Thought," *Comment*, 1964, republished at Anarchy Archives. http://dwardmac.pitzer.edu/anarchist_archives/bookchin/ecologyandrev.html

186   Murray Bookchin, "What Is Social Ecology?" *Environmental Philosophy: From Animal Rights to Radical Ecology*, ed. M.E. Zimmerman, Englewood Cliffs, NJ: Prentice Hall, 1993. Republished at Anarchy Archives. http://dwardmac.pitzer.edu/Anarchist_Archives/bookchin/socecol.html

NAS

of the United States, gave sustainability one of its most powerful weapons in linking social inequities to environmental degradation. He introduced four "laws of ecology" in his 1971 book, *The Closing Circle*. His first law of ecology held that "Everything is connected to everything else." Commoner developed this "law" in part to explain why every piece of the environment was fragile and utterly dependent on all other parts, and he heeded his own message devotedly. He took up the habit of wearing wrinkled shirts to save the energy it would have required to iron them. But Commoner also expanded his law to incorporate social interconnectedness as well. He summed up in his book, *The Closing Circle*,

ECOLOGIST BARRY COMMONER
The Emerging Science of Survival

> To resolve the environmental crisis, we shall need to forgo, at last, the luxury of tolerating poverty, racial discrimination, and war. In our unwitting march toward ecological suicide, we have run out of options. Now that the bill for the environmental debt has been presented, our options have been reduced to two: either the rational, social organization of the use and distribution of the earth's resources, or a new barbarism.[187]

Scientific misconduct, abetted by freewheeling capitalism, became the danger that Commoner argued against most forcefully. He drew out the theme in a series of books: *Science and Survival* (1966), which cast science as an overgrown Frankenstein project that threatened mankind's own survival, *The Closing Circle* (1971), which indicted modern technology, and *The Poverty of Power* (1976), which commandeered the laws of thermodynamics to endorse Commoner's particular policy preferences. Commoner became a celebrity, his agenda beloved by many. *Time* put him on its cover in 1970. He garnered awards and at least eleven honorary degrees.

Power Shift

Sustainability's social and economic impulses paired with its environmentalist ethos make it a magnet for youth. Sustainability, like its predecessor movements that excited student passions, invokes moralistic duties to repair and restructure the Earth. It dangles warnings of impending deadlines and narrow windows of opportunity to effect change, and urges adherents to act now, before it is too late. It encourages exuberance and committal, and rewards its followers with a sense of belonging to a community of the enlightened few. It endows the smallest actions with meaning and significance; something as small as recycling a plastic cup and turning up the summer thermostat by one degree become noble sacrifices rewarded with laurels, per Commoner's maxim, that are connected to massive ecosystems and contribute

---

187   Cited in Glenn Ricketts, "The Roots of Sustainability," *Academic Questions*, Vol. 23 No. 1, Spring 2010.

inexorably towards the saving of the planet.

Diversity, with its demands for racial reconciliation, affirmative action, multicultural sympathies, and tokens of reparation, finds in sustainability a metanarrative that links its specific grievances to a larger circle of global oppression that must be smashed. Social justice finds justification for its communitarian fervor in sustainability's calls for a new economic and social order. Feminism fawns over sustainability's support for birth control and abortion, as well as its calls for gender equality and female empowerment. Economic reformers and socialist-sympathizers echo sustainability's disapproval of free markets.

The economic tie is one of the strongest in sustainability. Many sustainability critics have noticed this. Not least among these is Brian Sussman, a former meteorologist and now radio host in the San Francisco Bay Area. Sussman has chronicled many of the links between sustainability initiatives and economic reorganization in his books *Climategate* and *Eco-Tyranny*. Many sustainability advocates have acknowledged the link between sustainability and anti-capitalism as well. At a "Power Shift" rally for youth environmentalists in 2009, Van Jones, President Obama's former green jobs czar, shouted,

> *This movement is deeper than a solar panel, deeper than a solar panel. Don't stop there. Don't stop there. No, we're going to change the whole system. We're going to change the whole thing!*[188]

Naomi Klein, a Canadian-born activist and comrade of Bill McKibben's fossil free divestment campaign, has long advocated for economic revolution as the surest way to achieve environmental sustainability.[189] Her 2000 bestseller *No Logo* castigated consumers for brand loyalty and tore into corporations for embracing globalism. Her 2007 book *The Shock Doctrine: The Rise of Disaster Capitalism* shinnied up the *New York Times* bestseller list for its brazen accusation that Milton Friedman-esque capitalists were capitalizing on the chaos in disaster-ridden zones to stealthily set up free markets that undermined social safety nets and sharing economies. Klein's most recent work, *This Changes Everything: Capitalism Vs. the Climate*, released in September 2014, bluntly confesses that nothing but an overthrow of free market capitalism can save the environment. Her website admonishes, "Forget everything you think you know about global warming. The really inconvenient truth is that it's not about carbon—it's about capitalism."[190] Some participants in the sustainability movement are bothered by Klein's overly-frank attack on capitalism. They would prefer to undermine it slowly than to identify the movement explicitly with international socialism—but this is a question of mood and atmosphere, not substance.

---

188   Van Jones, Keynote Address, Power Shift Conference on Climate Change, Washington, D.C., February 27, 200.

189   See Introduction.

190   Naomi Klein, "Watch the Book Trailer for This Changes Everything: Capitalism vs the Climate," NaomiKlein.org, August 20, 2014.   http://www.naomiklein.org/articles/2014/08/watch-book-trailer-changes-everything-capitalism-vs-climate

Conservatives and Sustainability

Not all initiatives labeled "sustainability-related" involve managed economies, radical feminism, diversity, and social justice. Many who align more closely with conservative principles see the rise of sustainability as an opportunity to endorse small government, personal frugality, and a local agrarian work ethic.

On the whole, though, this divide in the movement is not deep. Proportionally more energy, money, and manpower is devoted to the more extreme versions of sustainability. Conservative sustainability is in many ways a reaction to radical sustainability, an attempt to reclaim the term and opportunistically to seize a buzzword for alternative ends.

Still, the conservative/radical split is rooted in the idea of sustainability itself. A little etymology might be clarifying.

Like most watchwords in an era—popularized and eagerly claimed and reclaimed and circumscribed by divergent factions—sustainability is complicated. To sustain something, literally, is to carry it, to hold it up in place, and, negatively, to keep it from falling down or giving way. "Support, maintain, uphold," the Oxford English Dictionary puts it.[191] The word carries an aura of solidness, integrity, soundness, as in a pillar holding up a vaulted cathedral ceiling, or a connotation of satisfaction and fullness, as in a hearty meal that provides strength for the day's work. There's an air, too, of responsibility and wisdom in taking care, enjoying the luxury of time and energy to think ahead, maintain, keep in good repair, and plan for thefuture.

Sustainability's most remote etymological ancestor, the classical Latin *sustinere*, to "hold up, support, to maintain, preserve," is virtually identical to the confident optimism of its contemporary English echo. Sustainability's end goals—preservation, maintenance, support—seem to be givens, not questions. So long as you maintain the pillars, protect the marble, and keep sledge hammers out, the pillars will not falter and the ceiling will not cave. Westminster Abbey, Salisbury Cathedral, and Winchester Cathedral still soar high even at their near-millennium ages.

But the word's earliest Anglo progenitor, dated by the OED to 880, the Anglo-Norman and Old French *sustenir*, is a grim verb meaning to "bear, withstand, or endure." It has the hard edges of adversity, and it conveys the idea of some difficulty to be tolerated and gotten through. *Sustenir* describes a man struggling to sustain, to hold up off of himself a heavy weight in order to keep it from falling and crushing him. He carries this load not because the thing ought to be lifted high and held up, or that it naturally

---

191   "Sustain," Oxford English Dictionary, Oxford University Press. http://www.oed.com/view/Entry/195209?rskey=Y7yN31&result=2#eid

should rest atop something else, but that somehow it must. Survival demands it.

The two roots reveal a contrast. Which is it? Is sustainability confident preparation for and wise maintenance of a grand inheritance, or a desperate struggle for mere existence?

To varying degrees, it's both. Sometimes sustainability takes on the slow Southern drawl of local agrarianism, a generational respect for the earth and the fruit it bears, and a deliberate caretaking to preserve it and to love it. Sometimes sustainability radiates out from a cutting edge Silicon Valley start-up that engineers better, faster, more efficient ways to produce more with less energy. At other times, the murmurs of sustainability whisper in an old-fashioned frugality reminiscent of a previous generation that wasted nothing, wanted nothing, and saved and reused everything from glass jelly jars to yesterday's newsprint.

The conservative case for sustainability is perhaps best articulated by Roger Scruton, who contends that although environmentalism "has all the hall-marks of a left-wing cause"—a focus on victims, a dislike of capitalism, a quasi-redemptive religious character, and an especially strong appeal to youth—environmentalism sits well with conservative principles. "Indeed," Scruton writes,

> environmentalism is the quintessential conservative cause, the most vivid instance in the world as we know it, of that partnership between the dead, the living and the unborn, which Burke defended as the conservativearchetype.[192]

Scruton then identifies free markets and property rights as the most efficient means of internalizing externalities such as pollution, and the love and loyalty to home as the deepest link to an ethic of environmental preservation.

Another source of sustainability inspiration for many conservatives is Wendell Berry, the Stanford-educated English professor who left a teaching position at New York University to teach creative writing at the University of Kentucky and to farm in his native Henry County. Berry's prolific writing often praises Jeffersonian yeoman virtues and the values of local, small-town communities. For instance, he writes in "Manifesto: The Mad Farmer LiberationFront,"

> Love the quick profit, the annual raise,
> vacation with pay. Want more
> of everything ready-made. Be afraid
> to know your neighbors and to die.

---

192 Roger Scruton, "Conservatism Means Conservation," *Modern Age*, Fall 2007, pp. 351-352.

NAS

And you will have a window in your head.

Not even your future will be a mystery

any more. Your mind will be punched in a card

and shut away in a little drawer.

... Give your approval to all you cannot

understand. Praise ignorance, for what man

has not encountered he has not destroyed....

Put your faith in the two inches of humus

that will build under the trees

every thousand years.

Listen to carrion – put your ear

close, and hear the faint chattering

of the songs that are to come.

Expect the end of the world. Laugh....[193]

The contrast of the "two inches of humus" slowly accumulating under the tree and "everything ready-made" that the profit-hungry crave highlights Berry's distaste for the life driven by the (unsustainable) taste for material wellbeing. That life is the reason man has "destroyed" everything that he knows how to operate and leads Berry to playfully advise, "Expect the end of the world." The poem appeared in the journal of the Context Institute, whose motto is "Whole-system pathways to a thriving sustainable future."

But the predominant sects of sustainability are harsher and brasher than Scruton's and Berry's conservative agrarianism. When politicians who worry about carbon levels, businessmen who seek better branding, and campus environmentalists who fear industrialism ask of a behavior, or a product, or an activity, *Is this sustainable?* they usually do not mean, *Is it itself able to be sustained and kept going?* That is a question about durability, not "sustainability" as the word is now used. What the politicians, businessmen, and campus activists are really asking is whether something contributes to the greater good of the environment. Behind this lies a vague concept of the Earth's wellbeing, an environmental harmony, and some kind of balance of trade with Nature, in which we *extract* and she *replenishes* in equal proportions.

Whether running shoes hold up for several hundred Saturday morning jogs, or paper towels can wipe an entire counter before tearing bears on their durability. Whether the shoes or the paper towels are sustainable depends on whether the rubber was purified and molded without emitting too much sulfur, and whether the paper had been recycled previously, and whether both had been manufactured and

---

193   Wendell Berry, "Manifesto: The Mad Farmer Liberation Front," *In Context*, Fall/Winter 1991.

NAS

transported locally in a hybrid biodiesel low-emissions truck.

The real concern is not whether the products will last a long time (though that is part of it) but whether they fit a standard thought to promote a particular way of making and buying goods. "Sustainable" is in this usage not really an adjective for an object's state of latent potentiality, but a noun. "Sustainable" and "sustainability" are labels for a way of life, a philosophy of thought, a political movement—a comprehensive worldview.

That worldview is the one that demands restructuring the economic system atop principles of equality rather than individuality; exchanging freedom and market systems for managed economies and bureaucratic oversight; demonizing industries deemed related to global warming (namely, fossil fuel companies); and calling for immediate shifts to cost-prohibitive "green" and "renewable" energy paid for by government subsidies. At times, sustainability gives up on technology altogether and shakes from its sandals the dust of modern society, vowing to return to primitive lifestyles at "harmony" with nature. Fearing overpopulation, sustainability calls for strict measures to reduce the numbers of human Earth-inhabitants. And sustainability views social and economic issues as intrinsically linked to environmental ones by a web of Western colonialism and male patriarchy; it absorbs into its folds the corollary movements for diversity, social justice, sexual liberation, and economic parity, making sustainability a catch-all for social grievances of all stripes.

This is the predominant version of sustainability that animates those who advocate for it. This is the sustainability that we critique here. We take no issue with conservation, pollution caps, recycling efforts, and attempts to instill responsibility and forethought. We like clean air and clean water, and we appreciate measures that effect both. We are all for a less consumption-focused, appetitive culture that preaches satisfaction through possessions, and for a return to simpler pleasures. But we are also advocates of economic sensibility, respect for individual liberty of action and of mind, and civil debate rather than dogmatic assertion. And while we are agnostic as to whether global warming is happening, and if so whether it stems from man-made causes (both of which are actively debated in the scientific literature), we advocate for principles of academic freedom that enable both sides to present evidence and be given a fair hearing. But the campus version of sustainability inhibits all of these, replacing them with doctrinaire declarations and enforcing the party line. A climate of intellectual freedom appears unsustainable in its wake.

Chronology

Figure 6. Timeline of Sustainability

- 1948 Fairfield Osborn releases *Our Plundered Planet*, launching a post-war revival of Malthusian thought.

- 1948 William Vogt publishes *Road to Survival*, warning that overpopulation taxed agricultural yields and led to environmental degradation.

- 1962 Murray Bookchin publishes *Our Synthetic Environment* under the pseudonym Lewis Herber.

- 1962 Rachel Carson publishes *Silent Spring*, blaming profit-driven businesses for using pesticides that she argued threatened to poison song birds.

- 1964 Lyndon B. Johnson signs the Wilderness Act.

- 1964 Murray Bookchin coins the term "social ecology" in "Ecology and Revolutionary Thought."

- 1968 Paul Erhlich, a Stanford entomologist, publishes his best-selling neo-Malthusian tract, *The Population Bomb*.

- 1969 UN creates the UN Fund for Population Activities, now the UN Population Fund.

- 1969 President Nixon signs the National Environmental Policy Act.

- 1970 President Nixon signs the Clean Air Act and creates the Environmental Protection Agency.

- 1970 University of Colorado Boulder establishes an environmental affairs office, and the University of California-Irvine establishes the nation's first "School of Social Ecology."

- 1970 The first Earth Day is held on April 22, with 20 million Americans participating.

- 1971 Barry Commoner publishes *The Closing Circle*, which spells out his four laws of ecology, including his well-known first law: "everything is connected to everything else."

- 1972 The Club of Rome publishes *Limits to Growth*, which predicts that within 100 years the earth's resources will be exhausted if population growth and resource extraction continue at their current rates.

- 1972 President Nixon signs the Clean Water Act.

- 1972 UN Stockholm Conference on the Human Environment holds the first of many meetings.

- 1973 President Nixon signs the Endangered Species Act.

- 1976 Lowell Ponte releases *The Cooling: Has the Next Ice Age Already Begun?*

NAS

- 1979 Three Mile Island nuclear power plant partially melts down in Dauphin County, Pennsylvania.

- 1979 The *Charney Report* warns of global warming.

- 1983 "World Commission on Environment and Development" chaired by Gro Harlem Brundtland is established by the United Nations General Assembly resolution 38/161 "Process of Preparation of the Environmental Perspective to the Year 2000 and Beyond."

- 1984 World Population Conference meets in Mexico City, where governments declare that "as a matter of urgency," family planning services should be available.

- 1987 World Commission on Environment and Development chaired by Gro Harlem Brundtland releases *Our Common Future* (the "Brundtland Report") that makes "sustainability" a familiar term.

- 1987 Edward B. Barbier introduces the Venn diagram depicting sustainability in "The Concept of Sustainable Economic Development," *Environmental Conservation* 14, no. 2.

- 1988 James Hansen, NASA scientist, testifies before Congress that global warming is happening.

- 1988 The Intergovernmental Panel on Climate Change (IPCC) is established by the UN.

- 1989 *Time* magazine's January 2 cover depicts Earth as "Planet of the Year."

- 1989 Bill McKibben writes *The End of Nature*.

- 1990 The Talloires Declaration introduces the first official commitment to environmental sustainability on the part of university administrators.

- 1991 Attendees at the People of Color Environmental Leadership Summit in Washington, D.C. issue a manifesto with 17 principles of Environmental Justice.

- 1992 The United Nations Conference on Environment and Development meets in Rio de Janeiro (the Earth Summit) and releases *Agenda 21*, a plan to achieve environmentally sustainable development by the twenty-first century.

- 1992 Al Gore publishes *Earth in the Balance* to make "the rescue of the Earth the central organizing principle for civilization."

- 1992 Al Gore introduces legislation to phase out chlorofluorocarbons, deemed the primary chemical causing an ozone hole over Antarctica and the Northern Hemisphere.

- 1992 The Union of Concerned Scientists issues "Warning to Humanity" about global warming.

- 1993 Second Nature is founded by Teresa Heinz and John Kerry to incorporate sustainability into all of higher education.

- 1999 Peggy Barlett launches the Piedmont Project to incorporate sustainability into non-environmental courses at Emory University.

- 2004 The movie *The Day After Tomorrow* stokes public fears over an ice age in the Northern Hemisphere.

- 2006 The Association for the Advancement of Sustainability in Higher Education   launches.

- 2006 Second Nature launches the American College and University Presidents' Climate Commitment with 12 founding signatories.

- 2006 Al Gore releases the documentary *An Inconvenient Truth* about the threat of global warming.

- 2007 International Sustainable Campus Network is   founded.

- 2009 Copenhagen Climate negotiations fail to result in any treaty or emissions agreement.

- 2012 Bill McKibben starts the national Go Fossil Free divestment campaign on college campuses.

Who's Who in Sustainability

Peggy Barlett is the Goodrich C. White Professor of Anthropology at Emory University, where she helped to start the Piedmont Project to infuse sustainability into non-environmental courses. Barlett is a proponent of the transcendent experience of "re-enchantment" with nature, which she describes as a way to "expand the scientific paradigm of an objective relationship with the natural world (based on science and the use of reason) to include a more personal reconnection with the living earth."

Murray Bookchin (1921-2006) presaged much of the sustainability movement by fusing together hitherto distinct ideas about social organization and the environment. Originally a Communist, then an anarchist, Bookchin worked as a labor organizer riling people about capitalism's contribution to the use of pesticides, pollution, and social discrimination. His 1962 book *Our Synthetic Environment* discussed the dangers of pesticides six months before Rachel Carson's *Silent Spring* appeared. Under the pseudonym Lewis Herber he also released *Small Is Beautiful*. Bookchin sparred with ecologists who wanted to radically limit population and saw environmentalism as a deeply pro-human position.

Gro Harlem Brundtland (1939-) was the guiding force behind the seminal document in the history of the sustainability movement, the UN World Commission on Environment and Development report *Our Common Future*, better known simply as the Brundtland Report. A medical doctor with an M.D. from the University of Oslo and Master of Public Health from Harvard, she worked in Norway's public health system for ten years before becoming Minister of the Environment and eventually three-time prime minister of Norway. She was director-general of the World Health Organization from 1998-2003 and is now a special envoy on climate change to Ban Ki-Moon, the secretary-general of the United Nations.

NAS

Rachel Carson (1907-1964) sparked a popular environmental movement by raising concerns that pesticides were harming wildlife. A science writer with a background in biology, she is best-known for her bestseller *Silent Spring* (1962), which depicted a future world in which song birds died after being exposed to DDT. *Silent Spring* followed her series of widely read books on environmental contamination: *Under the Sea Wind, The Sea Around Us,* and *The Edge of the Sea.*

Barry Commoner (1917-2012) gave sustainability one of its central adages, "everything is connected to everything else," in his book *The Closing Circle* (1971). Commoner was a zoologist and biologist who founded modern ecology. In 1970, the year of the first Earth Day, *Time* magazine put him on its cover and called him the "Paul Revere of ecology." A third-party candidate for president in 1980 and the founding director of the government-financed Center for the Biology of Natural Systems, Commoner blamed capitalism for many environmental problems and advocated the integration of social justice concerns about sexism and racism into environmental efforts. He publicly disagreed with Paul Ehrlich about the danger of rising global population.

Anthony D. Cortese (1947-) is a Senior Fellow of Second Nature. He was its co-founder along with John Kerry, Teresa Heinz, and Bruce Droste. He served as president from March 1993-August 2012. He was the organizer of the American College & University Presidents Climate Commitment and co-founder of the Association for the Advancement of Sustainability in Higher Education and the Higher Education Association Sustainability Consortium.

NAS

Paul Ehrlich (1932-) popularized the fear of escalating population and the resulting shortage of food and resources with his 1968 bestseller *The Population Bomb*. He subsequently lost a highly publicized bet with one of his critics, Julian Simon, a professor of business at the University of Maryland, over whether metals would become scarcer and costlier. In 1980 Ehrlich predicted (wrongly) that by 1990, five metals would become more expensive, Simon that all would decrease in price. Originally trained as an entomologist and appointed to a professorship in biology at Stanford in 1966, Ehrlich became Stanford's Bing professor of population studies in 1976. In 1990 he shared with E.O. Wilson the Crafoord Prize awarded by the Royal Swedish Academy of Sciences to recognize areas of science not covered by Nobel Prizes.

Teresa Heinz (1938-), born in Mozambique, is a philanthropist, the head of the Heinz family foundation, and the widow of Republican Senator John Heinz III. Married to John Kerry after her first husband's death in 1991, Heinz co-founded Second Nature with Kerry after attending the Earth Summit in 1992.

John Kerry (1943-) co-founded Second Nature with Teresa Heinz in 1993 to bring sustainability to American culture by way of colleges and universities, after attending the UN Earth Summit at Rio de Janeiro in 1992. As a Democratic Senator from Massachusetts for 28 years, Kerry won the title "Environmental Hero" from the League of Conservation Voters. He is now the U.S. Secretary of State.

NAS

Naomi Klein (1970-) is the author of *No Logo*, *The Shock Doctrine*, and *This Changes Everything*—best-sellers that accelerated anti-capitalism fervor within the sustainability movement. She is a contributing editor for *Harper's* and reporter for *Rolling Stone*, and writes a regular column for *The Nation* and *The Guardian* that is syndicated internationally by *The New York Times* Syndicate.

Bill McKibben (1960-) is an author and environmentalist whose 1989 book *The End of Nature* is often regarded as the first book for a general audience about climate change. He is the founder of 350.org, a global grassroots climate change movement, and of the Go Fossil Free divestment campaign now spreading at hundreds of college campuses. As Schumann Distinguished Scholar in Environmental Studies at Middlebury College, McKibben teaches students principles of environmental advocacy and sustainability. He is a fellow of the American Academy of Arts and Sciences, the 2013 winner of the Gandhi Prize and the Thomas Merton Prize, and the recipient of honorary degrees from 18 colleges and universities. *Foreign Policy* named him to their inaugural list of the world's 100 most important global thinkers, and the *Boston Globe* said he was "probably America's most important environmentalist."

Debra Rowe is one of the administrative architects of the sustainability movement. A professor of renewable energies and energy management at Oakland Community College for more than twenty years, she has made her mark in the sustainability movement by working in numerous higher education sustainability organizations and speaking at most sustainability conferences. She is the president of the U.S. Partnership for Education for Sustainable Development, a founder and facilitator for the Disciplinary Associations Network for Sustainability, an advisor to AASHE's Higher Education Associations Sustainability Consortium, and chair of the technical advisory committee for the Sustainability Education and Economic Development Resource Center. She is the editor of a 1,000 page anthology for high school and college students, *Achieving Sustainability: Visions, Principles, and Practices*.

NAS

CHAPTER 3: THE GLOBAL WARMING DEBATE

Today, the concept of sustainability is practically inseparable from the dangers of global warming or "climate change." The two ideas, however, have separate origins, and became entangled with one another at a particular time and place. We begin this chapter by tracing that history and then proceed to the debate on global warming itself.

As we explained in the introduction, this report takes no position on the existence of global warming or subsidiary issues, including its causes. The National Association of Scholars and the authors regard these as open questions best resolved by good scientific investigation, transparency, and debate. We recognize, however, that many advocates of the anthropogenic global warming hypothesis do not welcome such debate. They regard the matter as settled and view those who call for further debate as un-scientific or worse. The epithets "climate denier" and "denialist" are often deployed as invective against those who refuse to conform to the orthodoxy that AGW is as fully established as it needs to be for purposes of charting practicalpolicy.

In the eyes of those who strike this polemical position, our insistence on continued debate will inevitably be misinterpreted as an attempt to subvert well-established scientific facts. We regret that intransigence and view it as an obstacle to legitimate science. But we do not intend to let this with-us-or-against-us approach taken by many AGW supporters force us to abandon our commitment to open-mindedness. The view that AGW exists has some plausible science to support it, as does the view that AGW is an illusion. In this chapter we present in condensed form both sides of the argument. Some of the claims or the evidence in favor of or against the theory are weak. We take note of those too. Our general point is the urgent need for greater scrutiny of the premises of the theory and the nature of the evidence for and against it. Claims advanced merely on the *assumption* that the theory of global warming is valid rest on highly insecure foundations.

Could there be a campus sustainability movement in the absence of belief in an urgent anthropogenic global warming crisis? In principle, yes. There was a strong environmental movement long before the AGW hypothesis was invented, and the other components of the CSM—anti-capitalism and progressive social justice theory—are connected to AGW only by tenuous threads. Nonetheless the sustainability movement has bet heavily on the validity of AGW, and if the hypothesis proves false—or unsustainable— the movement would lose most of its credibility.

The high-stakes gamble that supporters of CSM have made on AGW thus warrants our attention. It is the presiding intellectual context of the movement, and one of the reasons that it has turned so frequently to

NAS

mistreatment of scientists who dissent from AGW. After reviewing the argument for and against AGW, we conclude the chapter with a survey of what has happened to some of the dissenters.

History

A conventional birth date for global warming as a scientifically-based projection of climate trends is 1979, when the National Academy of Sciences' Ad Hoc Study Group on Carbon Dioxide and Climate issued a report, "Carbon Dioxide and Climate: A Scientific Assessment," better known as the *Charney Report*.[194] Before the *Charney Report*, the main climate threat in public discussion was global cooling, and for a while the partisans of the idea that the earth is cooling and the partisans of the idea that the earth is warming overlapped with divergent predictions. In the early 1980s, the predicted catastrophe of global warming contended with the predicted catastrophe of "nuclear winter." In the decades before the*Charney Report*, the idea that had most scientific currency was that Earth faced the rising danger of another great glaciation, a return to the Ice Age that ended about 10,000 years   ago.

Newsweek ran a doomsday article in 1975, "The Cooling World," that cited a litany of scientists and concluded, "The central fact is that after three quarters of a century of extraordinarily mild conditions, the earth's climate seems to be cooling down."[195] That same year, *Science News* published "Climate Change: Chilling Possibilities" and quoted C.C. Wallen, the chief of the Special Environmental Applications Division at the World Meteorological Organization, who warned that the planet had been cooling since 1940 in such significant, consistent patterns that it was unlikely the pattern could reverse and revert to warmer temperatures.[196] Global cooling earned support from the director of climate research at the University of East Anglia,[197] the National Center for Atmospheric Research,[198] the CIA,[199] the National Oceanic and Atmospheric Administration,[200] and the National Academy of Sciences.[201]

The switch from "consensus" on global cooling to the idea that we are threatened instead by global warming began to gather momentum in the early 1980s. An October 1985 conference in Austria of the

---

194   National Academy of Sciences' Ad Hoc Study Group on Carbon Dioxide and Climate, *Carbon Dioxide and Climate: A Scientific Assessment*, 1979.

195   Peter Gwynne, "The Cooling World," *Newsweek*, April 28,   1975. http://denisdutton.com/newsweek_coolingworld.pdf

196   John H. Douglas, "Climate Change: Chilling Possibilities," *Science News*, Volume 107, March 1, 1975. https://www.sciencenews.org/sites/default/files/8983

197   AP, "There's a New Ice Age Coming!" *Windsor Star*, September 9, 1972. http://news.google.com/newspapers?id=lzl_AAAAIBAJ&sjid=PlEMAAAAIBAJ&pg=4365,2786655&dq=climate+expert+new+ice+age+coming+hubert+lamb&hl=en

198   Walter Orr Roberts, "A New World Climate Norm? Climate Change and its Effect on World Food," Aspen Institute for Humanistic Studies, and National Center for Atmospheric Research. http://www.iaea.org/Publications/Magazines/Bulletin/Bull165/16505796265.pdf

199   "A Study of Climatogical Research as it Pertains to Intelligence Problems," Office of Research and Development, Central Intelligence Agency, August 1974.   http://www.climatemonitor.it/wp-content/uploads/2009/12/1974.pdf

200   Gwynne, "The Cooling World."

201   National Academy of Science, *Understanding Climate Change: A Program for Action*, 1975.

NAS

World Meteorological Organization and International Council of Scientific Unions had announced that "greenhouse gasses" in the first half of the 21st century would see "a rise of global mean temperature … greater than any in man's history." And the conference took note of the increase "between 0.3°and 0.7°C" in global mean temperature in the previous one hundred years.[202] But global warming theory took off in earnest in 1988, the year after the Brundtland commission's report, *Our Common Future*, had introduced the concept of "sustainability" as a key social and political objective.

The key date is June 23, 1988. That's the day on which NASA scientist James Hansen testified to the Senate Energy and Natural Resources Committee that "The greenhouse effect has been detected and it is changing our climate now."[203]

Hansen's testimony was dramatic, and itcaptivated public attention. The drama was partly staged. Senator Timothy Wirth, who had organized the hearing, later boasted,

NASA scientist James Hansen testified to Congress about the greenhouse effect.

> *Believe it or not, we called the Weather Bureau and found out what historically was the hottest day of the summer. Well, it was June 6 or June 9 or whatever it was, so we scheduled the hearing that day, and bingo: It was the hottest day on record in Washington, or close to it. It was stiflingly hot that summer. [At] the same time you had this drought all across the country, so the linkage between the Hansen hearing and the drought became very intense.[204]*

He also arranged for staffers to open the windows the night before to let in the warm summer air, and to shut off the air conditioning. Wirth described the effect:

> *So Hansen's giving this testimony, you've got these television cameras back there heating up the room, and the air conditioning in the room didn't appear to work. So it was sort of a perfect collection of events that happened that day, with the wonderful Jim Hansen, who was wiping his brow at the witness table and giving this remarkable testimony.[205]*

202 See Rupert Darwall, *The Age of Global Warming: A History*, London: Quartet, 2013, pp. 98-107.

203 Andrew C. Revkin, "Years Later, Climatologist Renews His Call for Action," *New York Times*, June 23, 2008. http://www.nytimes.com/2008/06/23/science/earth/23climate.html?pagewanted=print

204 "Frontline Hot Politics Interviews Timothy Wirth," *PBS*, April 24, 2007. http://www.pbs.org/wgbh/pages/frontline/hotpolitics/interviews/wirth.html

205 Ibid.

Hansen, director of NASA's Goddard Institute for Space Studies, presented evidence that showed higher temperatures during the first five months of 1988 than during any other comparable period in the previous 130 years, since temperature records began. He said that NASA was 99 percent certain that the warming stemmed from artificial increases in atmospheric greenhouse gas levels, not from natural variation. "It is time to stop waffling so much and say that the evidence is pretty strong that the greenhouse effect is here," he told the *New York Times*.[206]

It was only a matter of four days after Hansen's declaration that the nation's leading industrial countries, the G-7, met in Toronto and also affirmed that "global climate change" was happening and required "priority attention."[207] Prime Minister Margaret Thatcher and President Ronald Reagan professed to be convinced— although President Reagan declined to support new environmental regulations that he believed might harm the American economy.

Immediately after the G-7 Toronto conference, Canadian Prime Minister Brian Mulroney convened a follow-on Toronto conference organized by Gro Brundtland. This was the point of fusion between the concept of "sustainability" as laid forth in the Brundtland commission report and the concept of global warming or "climate change." Later in 1988, the newly formed Intergovernmental Panel on Climate Change (IPCC) met for the first time in Geneva. The IPCC's first chairman, Bert Bolin, embodied the new perspective that fighting climate change, pursuing "sustainable development," and treating sustainability as worldwide economic, social, and political reform were all part of a single unified project.

We are now more than a quarter-century past these events but we continue to live in what economist and historian Rupert Darwall has christened, "The Age of Global Warming." It is an age in which some very doubtful guesses about climate change are thoroughly mixed up with some very aggressive economic and political ambitions.

The scientific cases for and against anthropogenic global warming are complex and intricate. Hundreds of scientists have devoted their careers to understanding how local climates operate, how worldwide phenomena arise and occur, and what spurs temperature changes. Many matters are highly contended in the scientific literature. Below, we summarize the arguments and evidence commonly summoned on both sides of the debate.

---

206   Philip Shabecoff, "Global Warming Has Begun, Expert Tells Senate," *New York Times*, June 24, 1988. http://www.nytimes.com/1988/06/24/us/global-warming-has-begun-expert-tells-senate.html

207   "Environment," *Toronto Economic Summit Economic Declaration*, G-7 Summit, June 21, 1988. http://www.g8.utoronto.ca/summit/1988toronto/communique/environment.html

NAS

Global Warming: Yes

Sustainability derives its force primarily from two issues: global warming and economic inequality.

Global warming is seen as an imminent, non-reversible, potentially lethal threat to humans and animals. Even mild warming endangers the steadily rising quality of life that people across the globe have been enjoying. According to the Intergovernmental Panel on Climate Change, an increase of anything more than 2 degrees Celsius will prove catastrophic. To keep countries on target, the IPCC created a "carbon budget" of 1 trillion tons that humans may burn before triggering a net 2-degree rise. If fossil fuel use continues at its present rate, that budget will be expended within 30 years.[208]

The effects of climate change manifest themselves in various ways. Warmer temperatures cause heat stroke and heat exhaustion. Shrinking glaciers dump water into increasingly full oceans, swallowing up acres of shoreline and the communities that live along them.

Warmer air holds more water than cold air, upping the ante on hurricanes and typhoons. Sandy, the category 3 hurricane in October 2012 that hit New York and New Jersey, wrecked Manhattan's Financial District and coastal areas of Queens and Brooklyn. Typhoon Haiyan in November 2013, with its 195 mile-per-hour winds, swept the Philippines even as UN climate change representatives convened in Warsaw. Haiyan left more than 6,000 Filipinos dead and another 4 million homeless or displaced. The Philippines' climate negotiator, Yeb Sano, delivered an impassioned speech to the delegates at Warsaw: "What my country is going through as a result of this extreme event is madness... To anyone outside who continues to deny and ignore the realities of climate change, I dare them, I dare them to get off their ivory towers and away from the comfort of their arm chairs. I dare them to go to the islands of the Pacific."[209]

The increasing water density of the air also makes for snowier winters. And by evaporating more water from the earth's surface, the warming strains desert flora and fauna. The resulting forest fires in the American southwest eat up historic mountain preserves and people's homes. Droughts in Africa lead to famines.

The problems intensify, though, when the accumulated effects of climate change trigger runaway global warming. Climate scientists warn that the earth is at the edge of three tipping points. The first involves melting snow and ice, which alter the earth's albedo effect, or the ability of the earth's surface to reflect sunlight. White snow bounces much of the sun's light back to space, but green, blue, and brown absorb

---

208 "Understanding the IPCC Reports," World Resources Institute. http://www.wri.org/ipcc-infographics

209 Yeb Sano, "'It's time to stop this madness' – Philippines Plea at UN Climate Talks," *Responding to Climate Change*, November 13, 2013. http://www.rtcc.org/2013/11/11/its-time-to-stop-this-madness-philippines-plea-at-un-climate-talks/

NAS

the sun's rays, leading to more warming, to more ice melting, and to more heat absorption.

The second is arctic methane gas fifty times more potent that CO2, currently frozen into the Tundra. As the ice melts, the gas is released, triggering more warming and more atmospheric   methane.

Ocean acidification, the third danger point, involves the health of the seas. As carbon dioxide sinks into the ocean and the water acidifies, plankton, the basis of the marine food chain, die and endanger all other aquatic life. These problems operate exponentially, and because the natural system features time lags,the full effects of today's decisions will not be felt for years to come.

Bill McKibben, one of the leaders of the environmental movement and founder of the advocacy group 350.org, broke down the data in his 2012 *Rolling Stone* article, "Global Warming's Terrifying New Math," which went viral and sparked a student campaign against the fossil fuel industry:

> *Meteorologists reported that this spring was the warmest ever recorded for our nation – in fact, it crushed the old record by so much that it represented the "largest temperature departure from average of any season on record." The same week, Saudi authorities reported that it had rained in Mecca despite a temperature of 109 degrees, the hottest downpour in the   planet's history.*

> *… So far, we've raised the average temperature of the planet just under 0.8 degrees Celsius, and that has caused far more damage than most scientists expected. (A third of summer sea ice in the Arctic is gone, the oceans are 30 percent more acidic, and since warm air holds more water vapor than cold, the atmosphere over the oceans is a shocking five percent wetter, loading the dice for devastating floods.) Given those impacts, in fact, many scientists have come to think that two degrees is far too lenient a target. "Any number much above one degree involves a gamble," writes Kerry Emanuel of MIT, a leading authority on hurricanes, "and the odds become less and less favorable as the temperature goes up." Thomas Lovejoy, once the World Bank's chief biodiversity adviser, puts it like this: "If we're seeing what we're seeing today at 0.8 degrees Celsius, two degrees is simply too much." NASA scientist James Hansen, the planet's most prominent climatologist, is even blunter: "The target that has been talked about in international negotiations for two degrees of warming is actually a prescription for long-term disaster." At the Copenhagen summit, a spokesman for small island nations warned that many would not survive a two-degree rise: "Some countries will flat-out disappear." When delegates from developing nations were warned that two degrees would represent a "suicide pact" for drought-stricken Africa, many of them started chanting, "One degree, one Africa." … The official position of planet Earth at the moment is that we can't raise the temperature more than two degrees Celsius – it's become the bottomest of*

NAS

bottom lines. Two degrees.[210]

Since McKibben wrote his piece—which has become a touchpoint for many in the movement—scientists have concluded that 2014 has surpassed previous records as the hottest year. With the ten warmest recorded years all occurring since 1997, this latest data point indicates a growing trend.[211] The *New York Times*, reporting on the rising temperature, put the news in historical context:

> *February 1985 was the last time global surface temperatures fell below the 20th-century average for a given month, meaning that no one younger than 30 has ever lived through a below-average month. The last full year that was colder than the 20th-century average was 1976.*[212]

The cause of such warming is largely attributed to greenhouse gas emissions. Pennsylvania State University climate scientist Michael Mann, quoted by the *Times*, noted that

> *It is exceptionally unlikely that we would be witnessing a record year of warmth, during a record-warm decade, during a several decades-long period of warmth that appears to be unrivaled for more than a thousand years, were it not for the rising levels of planet-warming gases produced by the burning of fossil fuels.*[213]

Global warming caused by increases in atmospheric carbon dioxide has happened before in the geologic history of Earth. Therefore it can happen again. Recently, for example, geologists concluded that a major increase in CO2 was responsible for Paleocene-Eocene thermal maximum: "About 55.5 million years ago, a burst of carbon dioxide raised Earth's temperature 5°C to 8°C, which had major impacts on numerous species of plants and wildlife."[214]

In addition to threatening human life and wellbeing in general, global warming also harms specific communities more than it harms others. Climate change is seen as exacerbating economic inequality by disproportionately visiting the injuries on the poorest, least-prepared.

Global warming caused primarily by Western industrialism and consumerism leads to flooding in Pacific

---

210 Bill McKibben, "Global Warming's Terrifying New Math," *Rolling Stone*, July 19, 2012. http://www.rollingstone.com/politics/news/global-warmings-terrifying-new-math-20120719

211 "NASA, NOAA Find 2014 Warmest Year in Modern Record," National Aeronautics and Space Administration, January 16, 2015. http://www.nasa.gov/press/2015/january/nasa-determines-2014-warmest-year-in-modern-record/#.VMApP9LF_03

212 Justin Gillis, "2014 Breaks Heat Record, Challenging Global Warming Skeptics," *New York Times*, January 16, 2015. http://www.nytimes.com/2015/01/17/science/earth/2014-was-hottest-year-on-record-surpassing-2010.html?_r=0

213 *Ibid.*

214 Tim Wogan, "Greenhouse Emissions Similar to Today's May Have Triggered Massive Temperature Rise in Earth's Past," *Science*, December 15, 2014. http://news.sciencemag.org/climate/2014/12/greenhouse-emissions-similar-today-s-may-have-triggered-massive-temperature-rise?utm_campaign=email-news-latest&utm_source=eloqua

NAS

islands already beset by weak economies. Droughts, famines, and heat waves hit Africa especially hard, adding insult to colonial injury. Poor neighborhoods (often Black and Hispanic) in the United States are more likely to be situated near landfills and trash collection centers, or plants and factories that spew toxins. Minorities are more likely to work dirty, dangerous jobs such as coal mining. And when natural disasters strike, they're least likely to receive aid and to recover quickly; when Sandy hit New York, Wall Street was drained, rebuilt, and fortified within a few weeks, while more than two years after the storm, Queens and Brooklyn residents are still resetting their lives.

Because those with lower socio-economic statuses have less mobility, they can't relocate to better areas. And because they have little capital and few political connections, they have a more difficult time altering government policy. In times like these, the people have no option left but to take to the streets, as they did in September worldwide at the People's Climate March. The Natural Resources Defense Council explains,

> Championed primarily by African-Americans, Latinos, Asians and Pacific Islanders and Native Americans, the environmental justice movement addresses a statistical fact: people who live, work and play in America's most polluted environments are commonly people of color and the poor. Environmental justice advocates have shown that this is no accident. Communities of color, which are often poor, are routinely targeted to host facilities that have negative environmental impacts— say, a landfill, dirty industrial plant or truck depot. The statistics provide clear evidence of what the movement rightly calls "environmental racism." Communities of color have been battling this injustice for decades.[215]

The solution, then, is to implement environmental justice. The EPA defines the term as "the fair treatment and meaningful involvement of all people regardless of race, color, national origin, or income with respect to the development, implementation, and enforcement of environmental laws, regulations, and policies."[216] An environmentally just world not only stops climate change and protects natural resources. It also ensures that those resources are distributed evenly among all people, that political systems do not privilege the well-educated and well-connected, and that entrenched social customs do not hold back certain identity groups. Ideally, natural resources are used in a sustainable way—that is, by using only what nature can replenish. But economic resources, too, are rationed and distributed equally, so that no one can hoard wealth or prevent the lowest rungs of society from climbing the economic ladder.

215   Renee Skelton and Vernice Miller, "The Environmental Justice Movement," Natural Resources Defense Council, October 12, 2006. http://www.nrdc.org/ej/history/hej.asp

216   "Environmental Justice Program and Civil Rights," Environmental Protection Agency, New England's Office of Civil Rights and Urban Affairs. http://www.epa.gov/region1/ej/

NAS

As a result of sustainable resources use and a sustainable economic system, social systems should also emerge in a manner that sustains human dignity and establishes the equality of human beings. To get to this better world, policies that uplift underprivileged groups such as women, racial minorities, the disabled, and those who identify as gay, lesbian, or bisexual require special consideration. Contraception, abortion, policies to close wage gaps, the legal recognition of gay marriage, affirmative action, and other social measures are thus linked with social sustainability. Social repression, it suggests, mimics environmental repression.

The International Society of Sustainability Professionals publishes a short guide, "Confused About Social Sustainability?" that spells out the basics of social sustainability initiatives in developed countries. Among its recommendations are curbing use of minerals and resources (such as many involved in cell phones and other electronics) that are associated with guerrilla conflicts in places like the Congo; holding businesses responsible for the outcomes associated with their products (i.e., rejecting the National Rifle Association's slogan that it's people, not guns, that kill); awarding jobs on the basis of social justice criteria to prioritize, for instance, the homeless; and offering free on-site day care to employees' children.[217]

Anthony Cortese, former president of Second Nature, spelled out what this ideal of sustainability might look like:

> *Imagine a society in which all present and future humans are healthy and have their basic needs met. What if everyone had fair and equitable access to the Earth's resources, a decent quality of life, and celebrated cultural diversity?*[218]

Thus old-school conservation, technology-based solutions to global warming, and individual choices to consume less are insufficient to achieve sustainability. A sustainable world involves social and economic shifts as well.

Global Warming: No

Others are not so sure—that global warming is happening, that man is causing it, that the warming is significant enough to be dangerous, that social and economic woes are causatively linked to environmental ones, or that sustainability is the right agenda for higher education. Our critique in this book focuses on the last two. But in fair-mindedness, here is the case against dangerous anthropogenic global warming. There are many distinct cases against the existence of global warming, man's role in causing it, and

217   Darcy Hitchcock and Marsha Willard, "Confused About Social Sustainability? What It Means for Organizations in Developed Countries," International Society of Sustainability Professionals. https://www.sustainabilityprofessionals.org/sites/default/files/Confused%20about%20social%20sustainability_0.pdf

218   Anthony Cortese, "The Critical Role of Higher Education in Creating a Sustainable Future," *Planning for Higher Education*, March-May 2003, pg. 15. http://www.aashe.org/resources/pdf/Cortese_PHE.pdf

NAS

the need to urgently stop it. In general, the skeptics hesitate to extrapolate long-term predictions from inherently volatile weather conditions. The earth's global mean temperature has defied the projections that the IPCC has set down in its periodic assessment reports, and in its most recent report, the IPCC was forced to quietly lower its predictions. In the release of its Fifth Assessment Report in 2013, the IPCC predicted a temperature increase of 0.3 to 4.8 degrees Celsius, down from 1.1 to 6.4 degrees Celsius in the Fourth Assessment Report (2007).

Still, surface temperatures have registered below even the lower bound of the IPCC's 2013 projection. Nor is it clear that recent temperatures have been skyrocketing or that 2014 was exceptionally hot. Many data sets indicate temperatures stabilizing and flattening since 1998. Three main research centers compile the global average surface temperature in real time: NOAA, the UK Met Office Hadley Centre, and NASA's Goddard Institute for Space Studies.[219] All three show 2014 temperatures very close to those in 2010, the warmest year on record. The differences are small enough that they are not statistically significant. NASA's dataset shows 2014 as 0.02 degrees Fahrenheit warmer than 2010—well within the margin of error of plus or minus 1 degree Fahrenheit.[220]

Other research groups measure the temperature of the lower atmosphere, rather than the earth's surface, in an attempt to get a more even record undisturbed by the urban heat island effect. Data from atmospheric readings indicate that the earth is several tenths of a degree cooler than 1998, the hottest year on record since these measurements began in 1979. In December 2014, both the Remote Sensing Systems and the University of Alabama-Huntsville datasets showed 2014 in line to rank among the top five warmest years, but not as the warmest one of all.[221]

So where's the missing heat? Pro-global warming scientists have come up with a lengthy series of theories to explain the recent decline in temperatures, but none has been conclusive. The current theory physicists are flirting with is the idea that the Atlantic Ocean is holding the heat and preventing it from warming up the surface of the land.[222] Eventually the oceans will get too hot, though, and the rest of the globe will start warming again. Before that theory took hold, scientists thought it might be the Pacific harboring the

---

219 Two other groups also compile surface temperature data, but one, the Berkeley Earth group, does not update in real time and so does not have data to tell whether 2014 is hottest or not. And the other, developed by Kevin Cowtan and Robert Way from the University of York, uses a contested method to "fill in" missing data.

220 David Whitehouse, "2014 Global Temperature Stalls Another Year," Global Warming Policy Foundation, January 16, 2015. http://www.thegwpf.com/2014-global-temperature-stalls-another-year/

221 Paul C. "Chip" Knappenberger and Patrick J. Michaels, "Current Wisdom: Record Global Temperature—Conflicting Reports, Contrasting Implications," *Cato at Liberty*, December 10, 2014. http://www.cato.org/blog/current-wisdom-record-global-temperature-conflicting-reports-contrasting-implications

222 Jane J. Lee, "Has the Atlantic Ocean Stalled Global Warming?" *National Geographic*, August 21, 2014. http://news.nationalgeographic.com/news/2014/08/140821-global-warming-hiatus-climate-change-ocean-science/

NAS

heat, but that turned out to be false.[223] Other theories blame China's increased use of coal[224] and the recent rise in volcanoes,[225] both of which emit sulfur dioxide into the atmosphere that bounces back the sun's rays; the success of the 1988 Montreal Protocol in banning chlorofluorocarbons, which deplete the ozone layer;[226] and declines in atmospheric water vapor, which acts in tangent with greenhouse gases to trap heat.[227]

There are other problems. The data is sketchy, and there are historical gaps in the records. Often climatologists have to piece together records from thermometer readings and proxy measures, such as samples of ice cores and tree rings, or else simply guess, in order to produce historical temperature records. And the means of recording current temperatures are often unreliable. Anthony Watts, a veteran broadcast meteorologist, found during a 2009 examination of temperature stations across the country that 89 percent were poorly situated—often next to exhaust fans of air conditioning units or amidst dark asphalt parking lots—and failed the National Weather Service's siting requirements. The margin of error on their temperature readings, calculated by the U.S. government, was between 2 and 5 degrees Celsius. The average surface warming that the IPCC had calculated, in part on the basis of these stations, was 0.7 degrees Celsius over the prior 50 years—significantly less than the margin of error.[228]

The widely-circulated figure that 97 percent of all scientists believe global warming is dangerous and man-made also has been discredited. The survey that produced that statistic misclassified some global warming skeptics as proponents by wrongly labeling some of their research. It also counted those who believe that at least "some" global warming comes from human influence as supporters of the view that man-made global warming is dangerous.[229] In fact, a number of scientists recognize mild global warming and attribute it to the increased human use of carbon-based fuels, but consider the current warming a net benefit rather than precursor to a greater harm.

---

223 Ben Jervey, "Where Global Warming Went: Into the Pacific," *National Geographic*, February 11, 2014. http://news.nationalgeographic.com/news/2014/02/140211-global-warming-pause-trade-winds-pacific-science-climate/

224 Richard Black, "Global Warming Lull Down to China's Coal Growth," *BBC*, July 5, 2011. http://www.bbc.co.uk/news/science-environment-14002264

225 "Volcanic Aerosols, not Pollutants, Tamped Down Recent Earth Warming, Says CU Study," *Be Boulder*, March 1, 2013. http://www.colorado.edu/news/releases/2013/03/01/volcanic-aerosols-not-pollutants-tamped-down-recent-earth-warming-says-cu

226 Francisco Estrada, Pierre Perron, Benjamín Martínez-López, "Statistically Derived Contributions of Diverse Human Influences to Twentieth-Century Temperature Changes," *Nature Geoscience*, 2013, Volume 6, pp. 1050–1055. http://www.nature.com/ngeo/journal/v6/n12/full/ngeo1999.html

227 Susan Solomon, Karen H. Rosenlof, Robert W. Portmann, John S. Daniel, Sean M. Davis, Todd J. Sanford, Gian-Kasper Plattner. "Contributions of Stratospheric Water Vapor to Decadal Changes in the Rate of Global Warming," *Science*, March 5, 2010: Vol. 327 no. 5970 pp. 1219-1223. http://www.sciencemag.org/content/327/5970/1219.abstract

228 Anthony Watts, "Is the U.S. Surface Temperature Record Reliable?" *The Heartland Institute*, 2009. http://heartland.org/sites/default/files/SurfaceStations.pdf

229 James Taylor, "Global Warming Alarmists Caught Doctoring '97-Percent Consensus' Claims," *Forbes*, May 30, 2013. http://www.forbes.com/sites/jamestaylor/2013/05/30/global-warming-alarmists-caught-doctoring-97-percent-consensus-claims/

NAS

While it is true that today's global surface temperature and lower atmospheric temperature are both slightly warmer than they were fifty years ago, the increase is mild and unlikely to continue much further. The Earth experiences climate cycles regularly. For instance, the Medieval Warm Period, from about 950 to 1300 AD, dramatically warmed the earth—so much so that there are records of thriving lush farms in Greenland.

In fact, moderate warming may actually benefit the earth. Warmer temperatures and increased concentrations of carbon stimulate lush plant growth, while mild weather (as opposed to historically frigid eras) benefits human wellbeing. Bjorn Lomborg, the Danish environmental economist and founder of the Copenhagen Consensus, recognizes the existence of global warming but discounts its harms. He calculates that globally warmer temperatures could actually save as many as 1.4 million lives per year.[230] Princeton physicist and former director of energy research at the Department of Energy Will Happer testified to Congress in 2009, "I believe that the increase of CO2 is not a cause for alarm and will be good for mankind, for among other reasons because of its beneficial effects on plant growth."[231] Happer wrote in the *Wall Street Journal* in March 2012,

> *CO2 is not a pollutant. Life on earth flourished for hundreds of millions of years at much higher CO2 levels than we see today. Increasing CO2 levels will be a net benefit because cultivated plants grow better and are more resistant to drought at higher CO2 levels, and because warming and other supposedly harmful effects of CO2 have been greatly exaggerated. Nations with affordable energy from fossil fuels are more prosperous and healthy than those without.*
>
> *The direct warming due to doubling CO2 levels in the atmosphere can be calculated to cause a warming of about one degree Celsius. The IPCC computer models predict a much larger warming, three degrees Celsius or even more, because they assume changes in water vapor or clouds that supposedly amplify the direct warming from CO2. Many lines of observational evidence suggest that this "positive feedback" also has been greatly exaggerated.[232]*

It's also unclear how much warming is due to human influence. Richard S. Lindzen, an MIT professor of meteorology, commented in the *Wall Street Journal*,

> *The main statement publicized after the last IPCC Scientific Assessment two years ago was that*

---

230 Bjorn Lomborg, "Global Warming Will Save Millions of Lives," *The Telegraph*, March 12, 2009. http://www.telegraph.co.uk/comment/personal-view/4981028/Global-warming-will-save-millions-of-lives.html

231 William Happer, "Climate Change: Statement of William Happer, Cyrus Fogg Brackett Professor of Physics, Princeton University, Before the U.S. Senate Environment and Public Works Committee, Senator Barbara Boxer, Chair," U.S. Senate Environment and Public Works Committee, February 25, 2009. http://www.epw.senate.gov/public/index.cfm?FuseAction=Files.View&FileStore_id=84462e2d-6bff-4983-a574-31f5ae8e8a42

232 William Happer, "Global Warming Models Are Wrong Again," *Wall Street Journal*, March 27, 2012. http://online.wsj.com/articles/SB10001424052702304636404577291352882984274

it was likely that most of the warming since 1957 (a point of anomalous cold) was due to man. This claim was based on the weak argument that the current models used by the IPCC couldn't reproduce the warming from about 1978 to 1998 without some forcing, and that the only forcing that they could think of was man. Even this argument assumes that these models adequately deal with natural internal variability—that is, such naturally occurring cycles as El Niño, the Pacific Decadal Oscillation, the Atlantic Multidecadal Oscillation, etc.

Yet articles from major modeling centers acknowledged that the failure of these models to anticipate the absence of warming for the past dozen years was due to the failure of these models to account for this natural internal variability. Thus even the basis for the weak IPCC argument for anthropogenic climate change was shown to be false.[233]

The historical records show many periods of warming and cooling, many of them so ancient that it is unlikely man even had the technological capacity at the time to be responsible for them. And there is evidence that global temperature swings are caused by sun spots, changes in the sun's electromagnetic activity because of variations in the intensity of solar wind, and the power of El Niño, which suppresses the cold upwelling off of South America. Indeed, 1998, one of the warmest years on record, saw one of the largest El Niños in recent history. Even if the earth may be warming, it's not certain that the warming will continue, or that it will become dangerous.

Climategate

One cause for skepticism of anthropogenic global warming is because of high-profile scandals in the field of climatology. One of the best-known, "Climategate," implicated some of the world's top climate scientists in a plan to keep out of the IPCC's publications any article skeptical of global warming, "even if we have to re-define what the peer-review literature is!"[234] They also worked together to selectively cull data that told the right story and modify or leave out data that did not. One of the most famous graphs implicated in Climategate was the "hockey stick" developed by University of Virginia climatologist Michael Mann that showed centuries of flat temperatures followed by rapidly increasing temperatures in the 20th century.

On November 19, 2009, just prior to the UN Climate Conference in Copenhagen, and again on November 22, 2011, before the UN Climate Conference in Durban, South Africa, several thousand emails involving top climate scientists from the United States and United Kingdom were posted online (whether they

---

233   Richard S. Lindzen, "The Climate Science Isn't Settled" *Wall Street Journal*, November 30, 2009. http://online.wsj.com/articles/SB10001424052748703939404574567423917025400

234   James Delingpole, "Climategate 2.0," *Wall Street Journal*, November 28, 2011. http://online.wsj.com/articles/SB1000142405297020445210457705983062600226

NAS

were leaked or hacked is still unknown). The email threads involved Mann and UK-based researchers at the Climatic Research Unit at East Anglia University who wrote the core of the IPCC's reports. When their independent research showed conflicting temperature graphs, they struggled with how to present their data. In a 1999 email, Mann wrote to his colleagues,

> *Keith's series...differs in large part in exactly the opposite direction that Phil's does from ours. This is the problem we all picked up on (everyone in the room at IPCC was in agreement that this was a problem and a potential distraction/detraction from the reasonably consensus viewpoint we'd like to show w/ the Jones e al and Mann et al series).*[235]

Keith Briffa, the climatologist at the Climatic Research Unit whose tree-ring data showed declining temperatures since 1960, wrote,

> *I know there is pressure to present a nice tidy story as regards "apparent unprecedentedwarming in a thousand years or more in the proxy data" but in reality the situation is not quite so simple.*[236]

In the end they omitted some of the tree-ring proxy data showing temperature declines, and inflated other dissenting data. They also suppressed the Medieval Warm Period, the well-documented period of warm temperatures from about 900 to 1300 AD. Phil Jones, the director of the Climatic Research Unit, wrote to Mann and several others about his successful massaging of his data to reflect the "consensus" temperature charts:

> *I've just completed Mike's [Mann] Nature trick of adding in the real temps to each series for the last 20 years (i.e. from 1981 onwards) and from 1961 for Keith's to hide the decline.*[237]

Michael Mann's "trick" was to substitute thermometer data for proxy data and vice versa as necessary to produce the hockey stick-shaped graph, without noting these substitutions.

The Climategate emails followed and confirmed earlier doubts about Mann's data. As early as 2003, Canadian economist Ross McKitrick and mining executive Stephen McIntyre began requesting original surface temperature data from Mann and his colleagues at the Climatic Research Unit and scrutinizing the numbers they found. The results of their examination, published in the journal *Environment and Energy*, found "collation errors, unjustifiable truncation or extrapolation of source data, obsolete data, geographical location errors, incorrect calculation of principal components and other quality control

---

235 Email from Michael Mann to Keith Briffa, Chris Folland, Phil Jones, September 22, 1999. http://www.assassinationscience.com/climategate/1/FOIA/mail/0938018124.txt

236 Email from Keith Briffa to undisclosed recipients, September 22, 1999. http://www.assassinationscience.com/climategate/1/FOIA/mail/0938018124.txt

237 Email from Phil Jones to Ray Bradley, Michael Mann, Malcolm Hughes, November 16, 1999. http://lrak.net/emails.html

NAS

defects."[238] Reversing Mann's errors and updating his data, they found that the temperature of the 15th century (at the end of the Medieval Warm Period) was warmer than any period in the 20th century.

McIntrye and McKitrick found that Mann's computing model that synthesized different data series gave more weight to the handful of series that showed hockey stick shapes and depressed the weight of those that did not. The weighting of Mann's model was so strong that when McIntyre and McKitrick experimented with feeding random data into his model, they found the result was still the flat hockey stick handle followed by a sharply rising paddle.[239]

In response, Mann argued that McIntyre and McKitrick had used a faulty version of his data and had failed to replicate his computer modeling system. Mann supplied McIntyre and McKitrick with a corrected version of his climate data, which they found to be nearly identical to the first set. He declined to release his full computer model.

In response to the Climategate scandal, the Climatic Research Unit announced that it no longer had the original data.[240] The University of East Anglia appointed two investigations. One, a Scientific Appraisal Panel of six university academics and chaired by Lord Ronald Oxburgh, investigated for three weeks and then released a five-page report. It cleared the Climatic Research Unit of any charges of "deliberate scientific malpractice" but acknowledged the Unit was "slightly disorganised" and that it would benefit from "close collaboration with professional statisticians."[241] A second UEA report, The Independent Climate Change E-mails Review of five panelists under Sir Muir Russell, released a longer, 160-page report that found a "consistent pattern of failing to display the proper degree of openness" but no reason to doubt the scientists' "rigor and honesty" or to "undermine the conclusions of the IPCC assessments."

The House of Commons reviewed the incident as well. After five weeks it announced,

In the context of the sharing of data and methodologies, we consider that Professor Jones' actions were in line with common practice in the climate science community...We are content that the phrases such as 'trick' or 'hiding the decline' were colloquial terms used in private e-mails and the balance of evidence that we have seen does not suggest that Professor Jones was trying to subvert

238   Stephen McIntyre and Ross McKitrick, "Corrections to the Mann et. al.   (1998) Proxy Data Base and Northern Hemisphere Average Temperature Series," *Environment and Energy*, Vol. 14, No. 6, 2003, pg. 751. http://www.uoguelph.ca/~rmckitri/research/MM03.pdf

239   Ross McKitrick, "What Is the 'Hockey Stick' Debate About?" Presentation at the APEC Study Group, Australia, April 4, 2005. http://www.uoguelph.ca/~rmckitri/research/APEC-hockey.pdf

240   "Steve Goreham, *The Mad, Mad, Mad World of Climatism*, New Lenox Books, 2012, pg. 162.

241   "Report of the International Panel Set up by the University of East Anglia to Examine the Research of the Climatic Research Unit," April 12, 2010. http://www.uea.ac.uk/mac/comm/media/press/CRUstatements/SAP

NAS

the peer review process.[242]

Mann is now embroiled in a lawsuit with his critics Mark Steyn, *National Review*, and the Competitive Enterprise Institute—though he is the not the defendant but the accuser, charging them with libel. Mann is also suing in Canada skeptical scientist Tim Ball for alleged libel.

Gorism

The case for manmade global warming can be put forward, as we have indicated, as a set of interrelated scientific hypotheses. But it also can be put forward as a sensational narrative that has little to do with science. The sensational narrative deserves attention too, because it plays a large role in sustaining the sustainability movement.

When movie makers in the 1950s conjured the peril of an invasion of ants the size of houses (*Them!* 1954) or mind-controlling aliens (*Invaders from Mars*, 1953) it was easy for audiences to indulge a make-believe tale of disaster. But the line between scary fantasy and actual peril isn't always so clear. Al Gore's 2006 movie, *An Inconvenient Truth*, has been understood by many Americans as a documentary on the peril of the world-wide catastrophe of global warming. But the peril depicted in *An Inconvenient Truth* may be no more real than the danger posed by extra-large ants or mean Martians. Gore's documentary incorporated Hollywood special effects, including footage from the cli-fi movie *The Day After Tomorrow*. The haunting images of glaciers falling into the sea in Gore's film were actually computer-generated imagery from the movie, passed off as the real thing.[243]

"Cli-fi"—climate fiction—is indeed a new genre in books as well as movies. A mass audience has emerged that is eager to be entertained by stories of how humans, having used up the world's resources or having inadvertently unleashed catastrophic climate change, must cope with the consequences. Hollywood has contributed *Waterworld* (1995), *Elysium* (2013), *Wall-E* (2008), *Snowpiercer* (2014), and *Into the Storm* (2014). Some movies follow Gore's example by blurring the line between documentary and fiction. *The Age of Stupid* is set in 2055 but incorporates what is presented as "archival footage" (some of it real) from the present. The voiceover in the trailer explains the title:

> *Will people in the future call our time 'The Age of Stupid?' Or will humanity find a solution to the world's most urgent problem? You decide.*

The movie was released in 2009 during United Nations Climate Week with the aim of influencing the

---

242 "The Disclosure of Climate Data from the Climatic Research Unit at the University of East Anglia," House of Commons Science and Technology Committee, pg. 163. Mar. 24, 2010. http://www.publications.parliament.uk/pa/cm200910/cmselect/cmsctech/387/387i.pdf

243 Noel Sheppard, "Gore Used Fictional Video to Illustrate 'Inconvenient Truth,'" *Newsbuster*, April 22, 2008. http://newsbusters.org/blogs/noel-sheppard/2008/04/22/abc-s-20-20-gore-used-fictional-film-clip-inconvenient-truth

NAS

negotiations at the Copenhagen Summit.[244]

Cli-fi, of course, extends to books as well. One review of recent works in the genre mentioned *Far North* (2009) by Marcel Theroux (*Washington Post*: "first great cautionary fable of climate change"); *I'm With the Bears* (2011) by Mark Martin; *Back to the Garden* (2012) by Clara Hume; *The Healer* (2013) by Antti Tuomainen; *Odds Against Tomorrow* (2013) by Nathaniel Rich; *Solar* (2010) by Ian McEwan; and *Wild Ones* (2013) by Jon Mooallem.[245] Jason Mark, in his *New York Times* essay on cli-fi, also mentions Margaret Atwood's *MaddAddam* trilogy (2003, 2009, 2013) and Jennifer Egan's Pulitzer Prize-winning *A Visit from the Goon Squad* (2010).[246]

The importance of these cinematic and literary depictions of climate disaster lies in their feeding the mythology rather than the science of global warming. An artistic rendering doesn't invite a dispute over the facts. It presents the imagined world as given, and often with sufficient verisimilitude as to pull the viewer or reader along with the vision regardless of the facts.

In the broader cultural debate over global warming, fiction is a powerful ally of those who argue in favor of the hypothesis. On the side of the skeptics, it is possible to view one major movie, *Interstellar* (2014), as rejecting the global warming hypothesis. *Interstellar*, like these other movies, is about global catastrophe, but the catastrophe on offer is blight, not warming.

The global warming hypothesis advocates have as allies artists working in many other media as well. There are playwrights committed to the cause.[247] And painters, sculptors, performance artists, photographers, architects, and musicians who lend their talents. Al Gore's contribution was to lend this nascent movement a clear theme, and that theme has been endlessly elaborated. Andy Revkin's folk-style song, "Liberated Carbon" (2013) —"Liberate some carbon, baby, it's the American way!"—might be the movement's anthem.[248]

The idea that the world is imperiled by environmental catastrophe has been part of the fabric of popular culture at least since the 1950s, but the catastrophe is a perpetually receding horizon. A few months after

---

244   Scott Thill, "The Age of Stupid Gets Smart on Enviropocalypse," *Wired*, September 18, 2009. http://www.wired.com/2009/09/review-the-age-of-stupid-gets-smart-on-enviropocalypse

245   Rebecca Tuhus-Dubrow, "Cli-Fi: Birth of a Genre," *Dissent*, Summer 2013. http://www.dissentmagazine.org/article/cli-fi-birth-of-a-genre

246   Jason Mark, "Climate Fiction Fantasy: What 'Interstellar' and 'Snowpiercer' Got Wrong," *New York Times*, December 9, 2014. http://www.nytimes.com/2014/12/10/opinion/what-interstellar-and-snowpiercer-got-wrong.html?_r=0

247   Chantal Bilodeau, "Creating a List of Climate Change Plays," *Artists and Climate Change*, November 1, 2014. http://artistsandclimatechange.com/2014/11/01/creating-a-list-of-climate-change-plays/

248   Andrew Revkin, "Liberated Carbon," YouTube, September 24, 2013.   https://www.youtube.com/watch?v=pzZ_M4rnD48

the first Earth Day in 1970, the Canadian musician Neil Young released his album *After the Gold Rush*, in which the title song evoked an eco-apocalypse. Young warned:

> *Look at Mother Nature on the run*
>
> *In the nineteen seventies.*

Mother Nature, however, declined to run out, and as Young performed the song in repertoire, he changed the line to "the nineteen eighties," then "the twentieth century," and then "the twenty-first century."

The New Orthodoxy

In an academic environment that derides moral absolutes, preaches tolerance, and prides itself on objectivity, sustainability has managed to become the pervading campus dogma. Global warming, especially, enjoys a privileged position as a rare scientific theory abovedebate.

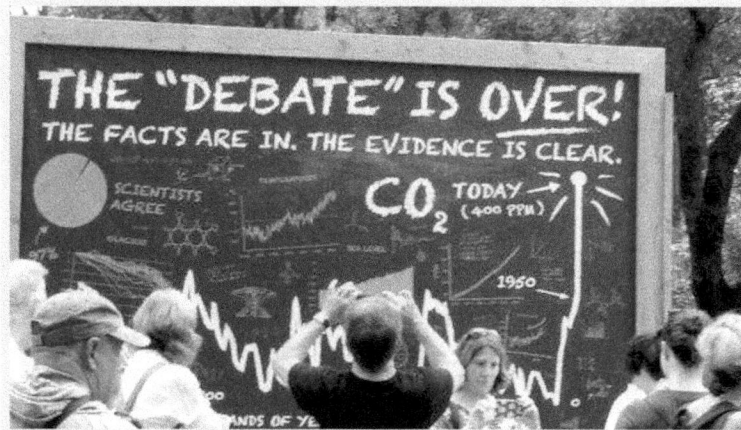

Scientists at the People's Climate March in New York City

Is the climate really changing? In the direction of global warming? Because of human activity? And if the answers to all these questions are "yes," are the interventions proposed by sustainability advocates plausible responses? These are key questions, but the sustainability movement does not welcome them. Instead it sets forth a set of doctrinaire answers and responds to virtually all challenges with declarations of "scientific" authority: that "consensus" has been achieved among all reputable observers and there is nothing left to be debated.

Consider the proclamations by some of AGW's most outspoken supporters.

Al Gore, interviewed on the CBS *Early Show* on May 31, 2006, told the public:

> *The debate among scientists is over. There is no more debate. We face a planetary emergency. There is no more scientific debate among serious people who've looked at the science...Well, I guess in some quarters, there's still a debate over whether the moon landing was staged in a movie lot in Arizona, or whether the earth is flat instead of round.*[249]

---

249 Rachel Waters, "Global Warming Movie Makes the Media Hot for Al Gore All Over Again," *Business and Media Institute*, Aug. 16, 2006, http://www.mrc.org/bmi/reports/2006/Summer_Rerun.html

NAS

At a June 2014 commencement at the University of California-Irvine, President Obama compared skeptics of anthropogenic global warming to those who might have told President Kennedy that the moon "was made of cheese." He told a sea of new graduates that "The climate change deniers suggest there's still a debate over the science. There is not."[250]

Secretary of State John Kerry echoed President Obama's theme in a 2014 speech on climate change:

> First and foremost, we should not allow a tiny minority of shoddy scientists and science and extreme ideologues to compete with scientific fact. ...I have to tell you, this is really not a normal kind of difference of opinion between people. Sometimes you can have a reasonable argument and a reasonable disagreement over an opinion you may have. This is not opinion. This is about facts. This is about science. The science is unequivocal. And those who refuse to believe it are simply burying their heads in the sand.[251]

It's not just politicians who speak so confidently. "The science is settled," then-chairman of the IPCC Robert Watson told an interlocutor during the 1997 Kyoto Protocol Treaty negotiations. "We're not going to reopen it here."[252]

Watson was succeeded in 2002 by Rajendra Pachauri, who commented in a 2008 interview with the *Chicago Tribune*,

> There is, even today, a Flat Earth Society that meets every year to say the Earth is flat. The science about climate change is very clear. There really is not room for doubt at this point.[253]

EPA chief Lisa Jackson testified in 2010 to Congress, "The science behind climate change is settled, and human activity is responsible for global warming."[254]

NASA has publicized the statistic that "Ninety-seven percent of climate scientists agree that climate-warming trends over the past century are very likely due to human activities" and keeps a running list of

---

250   Barack Obama, "Remarks by the President at University of California-Irvine Commencement Ceremony," The White House, Office of the Press Secretary, June 14, 2014. http://www.whitehouse.gov/the-press-office/2014/06/13/remarks-president-university-california-irvine-commencement-ceremony

251   John Kerry, "Remarks on Climate Change," Jakarta, Indonesia, U.S. Department of State, February 16, 2014. http://www.state.gov/secretary/remarks/2014/02/221704.htm

252   Henry Lamb, "Kyoto Report," *Eco-Logic*, Nov/Dec, 1997. http://sovereigntyonline.org/p/clim/kyotorpt.htm

253   Michael Hawthorne, "Blunt Answers About the Risks of Global Warming," *Chicago Tribune*, Aug. 3, 2008. http://articles.chicagotribune.com/2008-08-03/news/0808020393_1_global-warming-climate-change-rajendra-pachauri

254   Robin Bravender, "EPA Chief Goes Toe-To-Toe with Senate GOP Over Climate Science," *New York Times*. Feb. 23, 2010, http://www.nytimes.com/gwire/2010/02/23/23greenwire-epa-chief-goes-toe-to-toe-with-senate-gop-over-72892.html

NAS

scientific bodies that agree.[255] The American Physical Society declares,

> *The evidence is incontrovertible: Global warming is occurring. If no mitigating actions are taken, significant disruptions in the Earth's physical and ecological systems, social systems, security and human health are likely to occur. We must reduce emissions of greenhouse gases beginning now.*[256]

The American Association for the Advancement of Science has likewise proclaimed,

> *The scientific evidence is clear: global climate change caused by human activities is occurring now, and it is a growing threat to society.*[257]

But to other scientists, the evidence is anything but clear. A number of prominent scientists have expressed doubt that global warming is dangerous and man-made.

For instance, Richard Lindzen, emeritus Sloan Professor of Atmospheric Sciences at MIT, has strongly challenged much of the evidence cited to support anthropogenic global warming, arguing that the climate naturally varies, that the earth is recently emerging from a little ice age that lasted from the 15th to 19th centuries, that the artificial introduction of additional CO2 into the atmosphere is minor relative to the water vapor and clouds that dominate the atmosphere, and that the temperature data used to demonstrate dangerous warming are thin and easily nudged upwards. In a *Wall Street Journal* op-ed, "The Climate Science Isn't Settled," Lindzen writes,

> *The notion that complex climate "catastrophes" are simply a matter of the response of a single number, GATA (globally averaged temperature anomaly), to a single forcing, CO2 (or solar forcing for that matter), represents a gigantic step backward in the science of climate. Many disasters associated with warming are simply normal occurrences whose existence is falsely claimed to be evidence of warming. And all these examples involve phenomena that are dependent on the confluence of many factors.*[258]

William Happer, Princeton's Cyrus Fogg Brackett Professor of Physics and the former Director of Energy Research at the Department of Energy, holds that the climate, though mildly warming in recent years, poses no danger to human life. As he testified to the U.S. Senate in 2009, he has devoted his career "to understanding the interactions of visible and infrared radiation with gases—one of the main physical phenomena behind the greenhouse effect" and he sees no cause for alarm over the global temperature

---

255 "Consensus: 97% of Climate Scientists Agree," Global Climate Change: Vital Signs of the Planet, NASA. http://climate.nasa.gov/scientific-consensus/

256 *Ibid.*

257 *Ibid.*

258 Richard Lindzen, "The Climate Science Isn't Settled," *Wall Street Journal*, November 30, 2009. http://online.wsj.com/articles/SB10001424052748703939404574567423917025400

NAS

or the atmospheric carbon dioxide levels. Happer testified that

> *Without greenhouse warming, the earth would be much too cold to sustain its current abundance of life. However, at least 90% of greenhouse warming is due to water vapor and clouds. Carbon dioxide is a bit player. There is little argument in the scientific community that a direct effect of doubling the CO2 concentration will be a small increase of the earth's temperature—on the order of one degree. Additional increments of CO2 will cause relatively less direct warming because we already have so much CO2 in the atmosphere that it has blocked most of the infrared radiation that it can. It is like putting an additional ski hat on your head when you already have a nice warm one below it, but you are only wearing a windbreaker. To really get warmer, you need to add a warmer jacket. The IPCC thinks that this extra jacket is water vapor and clouds.*

> *Since most of the greenhouse effect for the earth is due to water vapor and clouds, added CO2 must substantially increase water's contribution to lead to the frightening scenarios that are bandied about. The buzz word here is that there is "positive feedback." With each passing year, experimental observations further undermine the claim of a large positive feedback from water. In fact, observations suggest that the feedback is close to zero and may even be negative. That is, water vapor and clouds may actually diminish the already small global warming expected from CO2, not amplify it.*[259]

Roy Spencer, the principal research scientist at the University of Alabama in Huntsville and US Science Team Leader for the Advanced Microwave Scanning Radiometer on NASA's Aqua satellite, testified to the Senate Environment and Public Works Committee that

> *My overall view of the influence of humans on climate is that we probably are having some influence, but it is impossible to know with any level of certainty how much influence. The difficulty in determining the human influence on climate arises from several sources: (1) weather and climate vary naturally, and by amounts that are not currently being exceeded; (2) global warming theory is just that – based upon theory; and (3) there is no unique fingerprint of human caused global warming.*[260]

Other distinguished scientists who are skeptical of dangerous anthropogenic global warming include Judith Curry, Professor of Earth and Atmospheric Sciences at the Georgia Institute of Technology; Freeman Dyson, Emeritus Professor of Physics at Princeton University's Institute for Advanced Study; S. Fred Singer,

---

259 William Happer, "Climate Change," Testimony before the U.S. Senate Environment and Public Works Committee, Senator Barbara Boxer, Chair, February 25, 2009. http://www.epw.senate.gov/public/index.cfm?FuseAction=Files.View&FileStore_id=84462e2d-6bff-4983-a574-31f5ae8e8a42

260 Roy Spencer, "Statement to the Environment and Public Works Committee of the United States Senate," July 18, 2013. http://www.epw.senate.gov/public/index.cfm?FuseAction=Files.View&FileStore_id=16e80c55-9ebf-42e4-852e-1f6e960b0902

NAS

Emeritus Professor of Environmental Sciences at the University of Virginia; Anthony Lupo, Professor of Atmospheric Science at the University of Missouri; and Ivar Giaever, Professor of Physicists at Rensselaer Polytechnic Institute and a 1973 Nobel laureate.

Another notable name is Bjorn Lomborg, the Danish intellectual famous for his strong data-driven research and his think tank, the Copenhagen Consensus Center. *The Guardian* named him one of the "50 people who could save the planet,"[261] and other publications, such as *Time, Esquire, Foreign Policy*, and *Foreign Policy & Prospect Magazine* have named him to their top-100 and top-75 lists of influential thinkers. Two of Lomborg's best-known books, *The Skeptical Environmentalist* (Cambridge University Press, 2001) and *Cool It* (Vintage Press, 2010), accept scientific data demonstrating warming temperatures but contend that the data do not support many claims about dangerous global warming, overpopulation, declining energy resources, deforestation, species loss, and other supposed consequences of global warming.

The oft-quoted "97 percent consensus" statistic has also been challenged as an Internet myth extrapolated from several incomplete surveys. Roy Spencer, along with Joseph Bast, president of the Heartland Institute, have scrutinized the number and found its sources wanting.[262] The main source, Spencer and Bast found, is from Maggie Kendall Zimmerman, a student at the University of Illinois, and Peter Doran, her master's thesis advisor, who published an article in *Eos Transactions American Geophysical Union*.[263] The article summarized the results of 79 responses by scientists to a two-question online survey. The questions asked whether mean global temperatures had generally risen since 1800, and whether human activity was a "significant contributing factor" in changing temperatures—questions that even scientists skeptical of dangerous anthropogenic global warming (who instead question the scale and the danger of the temperature change) would answer yes.

Another source for the "scientific consensus" is an essay by Harvard science historian Naomi Oreskes, who categorized the abstracts from 928 scientific articles published between 1993 and 2003. She calculated that 75 percent supported the "consensus view" that most of the warming over the last 50 years is caused by humans; that 25 percent took no position on anthropogenic climate change; and that "Remarkably, none of the papers disagreed with the consensus position."[264] In fact, Spencer and

261 "50 People Who Could Save the Planet," *The Guardian*, January 5, 2008. http://www.theguardian.com/environment/2008/jan/05/activists.ethicalliving

262 Joseph Bast and Roy Spencer, "The Myth of the Climate Change '97%,'" *Wall Street Journal*, May 26, 2014. http://online.wsj.com/articles/SB10001424052702303480304579578462813553136

263 Maggie Kendall Zimmerman and Peter Doran, "Examining the Scientific Consensus on Climate Change," *Eos Transactions American Geophysical Union*, Volume 90 No. 3, January 20, 2009, pp. 22-23. http://onlinelibrary.wiley.com/doi/10.1029/eost2009EO03/pdf

264 Naomi Oreskes, "The Scientific Consensus on Climate Change," *Science*, Vol. 306 no. 5702 December 3, 2004, pg. 1686. http://www.sciencemag.org/content/306/5702/1686.full

Bast found that Oreskes omitted numerous articles by scientists who consider themselves outside this "consensus."

William R.L. Anderegg, while a student at Stanford, examined 1,372 climate researchers whose work was listed on Google Scholar and found that of the 200 researchers "most actively publishing" work on the topic, "97-98%...support the tenets of ACC (anthropogenic climate change) outlined by the Intergovernmental Panel on Climate Change." Anderegg also concluded that "the relative climate expertise and scientific prominence of the researchers unconvinced of ACC are substantially below that of the convinced researchers."[265]   Anderegg's selection of the 200 "most active" of nearly 1,400 scholars left out most of the survey pool, and by concentrating on "anthropogenic climate change," he ignored the question of climate change's scale and danger—which is the more fiercely debated question concerning global warming.

The Perils of Dissent

Those who dare to question the "consensus" pay a high price for their intellectual openness. Scientists who persist in raising questions have been vilified. A special term of abuse, "denier," and its stronger alternative "denialist,"[266] modeled on the term "Holocaust denier," are frequently applied to such skeptics, along with a fair amount personal vitriol and ad hominem attack. In September 2014, when Steven Koonin, director of the Center for Urban Science and Progress at New York University and former undersecretary for science at the Department of Energy under President Obama, published an essay, "Climate Science Is Not Settled,"[267] *Time* editor Jeffrey Kluger responded with a scathing article denouncing "The Climate Deniers' Newest Argument."[268] In August 2013, environmentalist group 350 Action petitioned the World Meteorological Organization to name hurricanes after politicians who were "climate deniers."[269]

After the 400,000-strong People's Climate March in New York City, Robert F. Kennedy, Jr. called for climate skeptics to be jailed for their opinions.[270] James Hansen, the NASA scientist whose 1988 testimony to

265   William R. L. Anderegga, James W. Prall, Jacob Harold, and Stephen H. Schneider, "Expert Credibility in Climate Change," *Proceedings of the National Academies of Sciences*, April 9, 2010. http://www.pnas.org/content/107/27/12107.full

266   Ben Zimmer explains in the *Wall Street Journal* that "To call someone a 'climate-change denialist'—or, more succinctly, a 'climate denialist'—is a stronger accusation than simply calling that person a 'denier.' To be a denialist is to be caught in a reality-defying web of dogmatic 'denialism.' 'Denialism,' as Michael Specter of the *New Yorker* explained in his 2009 book with that title, is 'denial writ large—when an entire segment of society, often struggling with the trauma of change, turns away from reality in favor of a more comfortable lie.'" Ben Zimmer, "'Denialist': A Hot Epithet in Climate Rift," *Wall Street Journal*, September 27, 2014. http://online.wsj.com/articles/denialist-remains-a-popular-epithet-in-climate-battle-1411756770?mod=ST1

267   Steven Koonin, "Climate Science Is Not Settled," *Wall Street Journal*, September 19, 2014. http://online.wsj.com/articles/climate-science-is-not-settled-1411143565

268   Jeffrey Kluger, "The Climate Deniers' Newest Argument," *Time*, September 29, 2014. http://time.com/3445231/climate-denier-settled-science/

269   "350 Action Petitions the W.M.O. to Name Hurricanes After Actual Politicians who Deny Climate Change," 350 Action, August 26, 2013. http://350action.org/media/

270   Marc Morano, "Update: Video: Robert F. Kennedy Jr. Wants to Jail His Political Opponents – Accuses Koch Brothers of 'Treason'

NAS

the U.S. Senate catapulted the idea of global warming to public attention, holds that CEOs of fossil fuel companies "should be tried for high crimes against humanity and nature."[271] Lawrence Torcello, an assistant professor of philosophy at the Rochester Institute of Technology, argued similarly in a blog post at *The Conversation*:

> *The charge of criminal and moral negligence ought to extend to all activities of the climate deniers who receive funding as part of a sustained campaign to undermine the public's understanding of scientific consensus.*[272]

Major news organizations, including the BBC, have succumbed to the pressure to tell one-sided stories on climate change. In July 2014, after Nigel Lawson, Lord of Blaby, appeared on the "World at One" show to make the case that climate change had little to do with recent increases in flooding, activists complained to the BBC for permitting a dissenting voice. In response, the head of the BBC Complaints Unit, Fraser Steel, announced that "minority opinions and sceptical views should not be treated on an equal footing with the scientific consensus," and because Lawson's view "are not supported by the evidence from computer modelling and scientific research," the audience should have been warned of his aberrant status. Lawson has not been on the BBC since.[273] At a panel discussion at Columbia Journalism School in March 2015, *New York Times* environmental reporter Justin Gillis advised students that to present "deniers" as credible scientists would be to "perpetuate a lie" and violate journalism ethics. Covering both sides of the climate change story would mislead the public into believing that there was an actual debate on climate change, though of course "the facts are settled."

Activists have gone so far as to force the resignations of individuals as journal editors and members of commissions when they have broken ranks on the so-called "consensus." Others resign voluntarily when a skeptical viewpoint does get through the gatekeepers. When Roy Spencer and David D. Braswell published an article in the journal *Remote Sensing* suggesting that the computer models used to forecast climate change underestimated the atmosphere's ability to release energy into space, the journal's editor in chief, Wolfgang Wagner, resigned in protest.[274] Wagner faulted the peer review process for assigning three scientist reviewers who leaned towards skepticism of global warming, and criticized the paper

---

— 'They ought to be serving time for it,'" *Climate Depot*, September 21, 2014. http://www.climatedepot.com/2014/09/21/robert- f-kennedy-jr-wants-to-jail-his-political-opponents-accuses-koch-brothers-of-treason-they-ought-to-be-serving-time-for-it/

271   James Hansen, "Twenty Years Later: Tipping Points Near on Global Warming," *Huffington Post*, July 1, 2008. http://www.huffingtonpost.com/dr-james-hansen/twenty-years-later-tippin_b_108766.html

272   Lawrence Torcello, "Is Misinformation About the Climate Criminally Negligent?" *The Conversation*, March 13 2014. https://theconversation.com/is-misinformation-about-the-climate-criminally-negligent-23111#comment_333276

273   Raymond Snoddy, "BBC in Deep Water over Climate Change Censorship Row," *Newsline*, July 9, 2014. http://mediatel.co.uk/newsline/2014/07/09/bbc-in-deep-water-over-climate-change-censorship-row/

274   "Editor of *Remote Sensing* Resigns over Controversial Climate Paper; Co-author Stands by It," Retraction Watch, September 2, 2011. http://retractionwatch.com/2011/09/02/editor-of-remote-sensing-resigns-over-controversial-climate-paper-co-author-stands-by-it/

because "it essentially ignored the scientific arguments of its opponents"—though the entire paper focused on answering and discussing the arguments on the opposite side.[275] Spencer speculated on his blog that *It is obvious to many people what is going on behind the scenes. The next IPCC report (AR5) is now in preparation, and there is a bust-gut effort going on to make sure that either (1) no scientific papers get published which could get in the way of the IPCC's politically-motivated goals, or (2) any critical papers that DO get published are discredited with any and all means available.*[276]

Other scientists who depart from "consensus" have faced serious career setbacks, ad-hominem attacks, and scorn from their professional colleagues.

More Skeptics

Lennart Bengtsson is a senior research fellow at the Environmental Systems Science Centre at the UK University of Reading and the former director of the Max Planck Institute in Germany. A meteorologist, his scientific credentials are pristine. In 2005 he and several colleagues won the René Descartes Prize for Collaborative Research for their work on the Climate and Environmental Change in the Arctic project. In 2006 he received the 51st International Meteorological Organization Prize from the World Meteorological Organization.

Bengtsson is one of the foremost experts in climate science. Since 1990 he has expressed concerns about exaggeration and politicization in climate science, questioning the IPCC's predictions (which "should be taken with a grain of salt") and noting that the temperature over the Northern Hemisphere had been cooling for forty years.[277] In March 2014, he and several co-authors submitted a paper to *Environmental Research Letters*, one of the most prestigious journals, casting doubt on the IPCC's projected temperature rise of 2.0 to 4.5 degrees Celsius if the greenhouse gas emissions levels hold steady. Bengtsson and his colleagues suggested that the climate might not be as sensitive to carbon as the IPCC implied. Though Bengtsson is among the primary experts on these phenomena, the article was rejected. One reviewer privately commented that he rejected the paper was because it would hurt the climate change consensus: "Actually it is harmful as it opens the door for oversimplified claims of 'errors' and worse from the climate sceptics media side."[278]

---

275 Anthony Watts, "BREAKING: Editor-in-chief of Remote Sensing Resigns Over Spencer & Braswell Paper," *Watts Up With That*, September 2, 2011. http://wattsupwiththat.com/2011/09/02/breaking-editor-in-chief-of-remote-sensing-resigns-over-spencer-braswell-paper/

276 Roy Spencer, "More Thoughts on the War Being Waged Against Us," *Dr. Roy Spencer*, September 5, 2011. http://www.drroyspencer.com/2011/09/more-thoughts-on-the-war-being-waged-against-us/

277 Simon Rozendaal, "A Cool Blanket of Clouds," *Elsevier*, October 27, 1990. Reprinted and translated by Marcel Crok, "Bengtsson in 1990: 'One Cannot Oversell the Greenhouse Effect,'" *De Staat van het Klimaat*, May 13, 2014. http://www.staatvanhetklimaat.nl/2014/05/13/bengtsson-in-1990-one-cannot-oversell-the-greenhouse-effect/

278 Ben Webster, "Scientists in Cover-up of 'Damaging' Climate View," *London Times*, May 16, 2014. http://www.thetimes.co.uk/tto/science/article4091344.ece

NAS

Soon after, Bengtsson announced that he was joining the advisory council of the Global Warming Policy Foundation, an organization skeptical of anthropogenic global warming. Almost immediately he was forced to resign due to immense pressure from colleagues. In his resignation letter, Bengtsson wrote to the Global Warming Policy Foundation that he had received "such an enormous group pressure in recent days from all over the world that has become virtually unbearable to me," to the point of endangering his health and safety. He worried that

> *I see no limit and end to what will happen. It is a situation that reminds me about the time of McCarthy. I would never have expected anything similar in such an original peaceful community as meteorology. Apparently it has been transformed in recent years.*[279]

There are others as well. David Legates, a professor of climatology at the University of Delaware, and the Delaware State Climatologist from 2005 to 2011, dared to raise concerns about the data supporting stringent environmental regulations and got hit with an intrusive Freedom of Information Act (FOIA) request from Greenpeace. In December 2009, Greenpeace asked for all of his "e-mail correspondence and financial and conflict of interest disclosures...in the possession of or generated by the Office of the Delaware State Climatologist" from January 1, 2000, concerning "global climate change." Despite Delaware state law limiting FOIA requests to documents supported by public funding (which Legates did not receive as state climatologist) the vice president of the University of Delaware confiscated documents related to all of Legates's teaching, research, and service materials from 2000 to 2009, including work unrelated to the State Climate Office, whether conducted on Legates' own time or on university time, through his personal e-mail or his university e-mail, on his personal computer or a university computer, both in hard files and on computer disks. Ultimately the university decided to forego handing the documents over to Greenpeace, but only after several years of deliberation. Meanwhile, Legates was forced out of his positions as state climatologist, co-Director of the Delaware Environmental Observing System, and faculty advisor to the Student Chapter of the American Meteorological Society, and removed from all committee assignments within his department.

A few weeks after Greenpeace initially filed its FOIA request, however, the Competitive Enterprise Institute (a think tank skeptical of global warming alarmism) filed an identical request for information on the work of three University of Delaware professors who had contributed to the IPCC. This time the university vice president declined to collect any material from these professors. Legates, testifying before the U.S. Senate Committee on the Environment and Public Works, summed up his experience:

> *Scientists who deviate from the anthropogenic global warming playbook are likely to be harassed, have grants and proposals rejected without review, be treated more harshly than their peers, and*

---

279 Judith Curry, "Lennart Bengtsson Resigns from the GWPF," *Climate Etc.*, May 14, 2014. http://judithcurry.com/2014/05/14/lennart-bengtsson-resigns-from-the-gwpf/

be removed from positions of power and influence. I would have hoped that in the past decade, the discussion has become more civil. Indeed, a civil discussion can be had with some scientists that believe in the extreme scenarios of anthropogenic global warming. But too many in places of prominence and with loud voices have made this a war zone. Scientists like Bengtsson and myself have tenure or its equivalent and are somewhat insulated from the extreme attacks. But young scientists quickly learn to 'do what is expected of them' or at least remain quiet, lest they lose their career before it begins.[280]

This asymmetrical treatment of global warming skeptics and adherents is highlighted once more in the case of Michael Mann, who faced a FOIA request in 2013 from the American Tradition Institute and Virginia Delegate Robert Marshall related to his hockey stick graph. Ultimately the Virginia Supreme Court ruled that much of Mann's work need not be turned over. Before the decision was announced, the American Association of University Professors vigorously defended Mann's academic freedom, publishing statements on his behalf and filing an amicus brief in Mann's Supreme Court case. The brief held that

ATI's sweeping request, if allowed, would have a severe chilling effect on scientists and other scholars and researchers at public institutions of higher learning throughout the Commonwealth (and perhaps beyond). Put simply, Dr. Mann is a scientist and an academic, not a policymaker. And his unpublished research and internal communications with scientists are not part of any policy making function. If ATI is interested in how Dr. Mann's scholarship affects public policy, it should direct FOIA requests to the policymakers.[281]

When Legates had approached Joan DelFattore, president of the AAUP Chapter at the University of Delaware, about intervening in his case, however, DelFattore responded that FOIA matters "would not fall within the scope of the AAUP."[282]

James Enstrom, a University of California-Los Angeles epidemiologist, was dismissed from the UCLA School of Public Health, where he had taught for 34 years, after he popularized his research that suggests that the regulations of the California Air Resources Board were stricter than necessary. Enstrom had also alerted the public that the head of the board, Hier Tran, had bought a mail-order doctoral diploma showing a Ph.D. from the University of California-Davis, and that one of the board members, John Froines (one of his colleagues at UCLA) had served on the panel for 26 years, when the term limit is actually

---

280 David Legates, "Statement to the Environment and Public Works Committee of the United States Senate," June 3, 2014, pg. 24. http://www.epw.senate.gov/public/index.cfm?FuseAction=Files.View&FileStore_id=aa8f25be-f093-47b1-bb26-1eb4c4a23de2

281 Brief for the American Association of University Professors, pg 19, The American Tradition Institute and the Honorable Delegate Robert Marshall v. Rectors and Visitors of the University of Virginia and Michael E. Mann, 2013. http://www.aaup.org/sites/default/files/files/AAUP-amicus-brief-UVa%20-%20Va%20S%20Ct%20-%20FINAL(1).pdf

282 Legates, Statement, pg. 20.

NAS

three years. Tran had claimed that there were 2,000 premature deaths in California each year as a result of fine particulate pollution coming from diesel. Enstrom's research shows zero.[283] For his contribution to public knowledge about the risks of pollution and efforts to promote transparency within the Air Resources Board, Enstrom was rewarded with a humiliating termination of his career at UCLA.

These incidents of bias, censorship, and intellectual persecution come despite shaky scientific grounding for many of the theories of anthropogenic global warming behind which sustainability adherents hide. Besides the misconduct behind Michael Mann's "hockey stick" graph (which continues to be cited favorably), there is growing evidence that the EPA has broken legal peer review guidelines, soliciting reviews from employees of the very agencies that generated the reports under review, repeatedly (illegally) using the same reviewers who gave favorable reviews, and failing to publish scientific studies and regulations for public comment.[284] Both the Institute for Trade Standards and Sustainable Development and the Energy & Environment Legal Institute have found growing evidence to suspect the integrity and validity of the science behind the EPA's regulation of greenhouse gas emissions.[285]

We are *not* opposed to airing the theory of anthropogenic global warming, studying climate change, or improving the computer simulations on which much of contemporary climate science is based. But we do insist on an honest debate—a debate that climate change proponents often shy away from. And we insist on giving students the opportunity to consider and weigh the evidence, rather than swallowing ideological sound bites whole.

Such intolerance is especially hazardous for higher education, which depends more than any other institution on the give-and-take of open debate, transparency of evidence, civil exchange, and readiness to follow the best evidence to the most compelling conclusions. A form of education that is compromised by a rejection of these principles in favor of a doctrine backed by "consensus" is a form of education unsuited to a free society.

The Precautionary Principle

Advocates of the theory of manmade global warming have a fallback argument that they deploy when forced to deal with scientifically well-informed skeptics. The fallback is to admit that the evidence for large

---

283   Kelly Zhou, "UCLA Researcher James Enstrom not Reappointed to Position," *Daily Bruin*, August 30, 2010.

284   "New ITSSD FOIA Request Superseding Withdrawn FOIA Request No. EPA-HQ-2014-004938," Letter from Lawrence A. Koganto Mr. Larry F. Gottesman, National FOIA Officer, and Ms. Dana Hyland, U.S. Environmental Protection Agency, June 30, 2014. http://nebula.wsimg.com/e155ee64b03ea37237297cdbab7a2854?AccessKeyId=39A2DC689E4CA87C906D&disposition=0&al loworigin=1

285   Chris Horner, "Improper Collusion Between Environmental Pressure Groups and the Environmental Protection Agency as Revealed by Freedom of Information Act Requests," Energy and Environment Legal Institute, September 2014. http://eelegal.org/wp-content/uploads/2014/09/EE-Legal-FOIA-Collusion-Report-9-15-2014.pdf

NAS

impending changes in the earth's temperature may be weak or inconclusive but to observe that there is nonetheless a chance that the theory is right. If there is a reasonable probability that a catastrophe will befall the planet, doesn't it makes sense to act now to avert or at least mitigate the damage?

The term for this argument is "the precautionary principle." We examine it more closely later in this report. Here it will suffice to say that the precautionary principle is not an argument in favor of the global warming hypothesis, but a meta-argument that urges governments to take action *regardless* of the strength of the evidence for the theory.

The precautionary principle could depend on an assessment that the odds in favor of the theory being true are relatively high. If there is a 75 percent chance of a catastrophe, most people would say precaution is necessary. If the chance is one-in-a-million, most people would say the precaution is unwarranted. What are the odds that the manmade global warming hypothesis is true? The question is imponderable, since there are disputes about the basic facts.

In that sense, the precautionary principle seems more like a rhetorical maneuver than a clarifying idea.

The precautionary principle has been elaborated in other ways as well. Even if the chance of catastrophe is very low—say 1 out of a thousand—say some advocates of this idea, if the catastrophe is sufficiently dire, doesn't it make sense to act now to forestall it?

The advocates of "acting now" in the face of great uncertainty also sometimes introduce another argument beyond the precautionary principle: the idea that the cost of preventative measures will be more than off-set by the technological and social advances that striving for a carbon-free economy will induce. This idea lacks a succinct name, but it might be called "the pyramid principle," after the idea introduced by Humphrey Evans in 1979 that ancient Egypt's determination to build grandiose tombs for its pharaohs inadvertently spurred the rapid development of new technology and social organization. In the case of efforts to combat global warming, we can look forward to major advances in solar, wind, and other forms of technology, and perhaps more just and equitable conditions for all humanity.

The precautionary principle and the pyramid principle are, in some sense, unanswerable since they rely on pure speculation. But to the extent they amount to an actual economic rationale for global warming remediation, they have provoked at least one extended critique. Jim Manzi, an MIT-educated mathematician and founder and chairman of Applied Predictive Technologies, garnered public attention in a *National Review* cover article that argued that conservatives should accept the theory of global warming, but that "its impact over the next century could plausibly range from negligible to severe."[286] Manzi followed this article with several others in which he played out the idea of accepting at face value the IPCC's recommendations for remediating climate change. According to the IPCC's calculations at the time, the cost of climate change (if not prevented) would equal between "1 to 5 percent of global gross domestic product (GDP) sometime in the twenty-second century." Manzi notes that this seemingly small cost would in fact be "a huge amount of money," but that a much larger cost would come from efforts to forestall such climate change. This cost would come not only in direct expenses but also in the foregone wealth and benefits that would compound along the way. Manzi wrote at *The New Atlantis*, "Albert Einstein supposedly said that 'The most powerful force in the universe is compound interest' —and this mathematical reality is central to the wise evaluation of plans to address the risk of climate change."[287]

The precautionary principle, in other words may lead to a foolish lack of caution.

---

286   Jim Manzi, "Taking the Heat: A Conservative Strategy on Global Warming," *National Review*, June 25, 2007.

287   Jim Manzi, "Conservative, Climate Change, and the Carbon Tax," *The New Atlantis*, No. 21. Summer 2008, Pp. 15-25.

NAS

CHAPTER 4: HABITUATED: THE NUDGE-CULTURE OF SUSTAINABILITY

Yale president Peter Salovey recently gave his students some advice: "Fake it till you make it."[288]

The occasion for his fireside chat was the launch of Yale's new 2013-2016 Sustainability Strategic Plan.[289] On Wednesday, October 30, 2013, in the midst of the academic semester, Salovey gathered a group of students in the President's Room in Woolsey Hall and confronted them frankly with his goals for sustainability initiatives.

The previous three-year plan had set ambitious goals for energy and emissions reductions and increases in recycling.[290] Some of these Yale had met or slightly exceeded; others, it had failed by a long shot. The goal of 25 percent reduction in paper use, for example, flopped; Yale reduced its paper consumption by only 7 percent.[291]

Yale had evidently detected among its members a certain irresolution or perhaps a lack of heartfelt commitment to these policies. President Salovey's "faking it" solution was to make the students act as if they are enthusiastically eco-conscious, even if they are not. The idea is partly to prompt conformity to campus environmental policies. But the longer-term goal is that after a while, students won't be able to tell the difference between bluffing and believing. At some point, the mask begins to form the face.

In contrast to Yale's previous three-year sustainability plan, which had focused on institutional practices and policies, the new 2013-2016 plan deflects focus away from institutional environmental goals and towards personal behavioral changes. The plan isn't devoid of statistics and new policy targets: "Increase the renewable energy portfolio to represent 1 percent of the total electricity generated on campus by June 2016,"[292] "Establish sustainable procurement standards by June 2014,"[293] "Reduce sodium content in on-campus food offerings to 2,200 milligrams daily by June 2016."[294] But it does focus in a powerful way on personal behavior and private, everyday decisions. A section of the plan called "Leadership &

288   Hannah Schwarz, "Sustainability Plan Launched," *Yale Daily News*, October 31, 2013. http://yaledailynews.com/blog/2013/10/31/sustainability-plan-launched/

289   *Sustainability Strategic Plan 2013-2016*, Yale University, Yale Sustainability, August 2013. http://sustainability.yale.edu/sites/default/files/sustainabilitystrategicplan2013-16_0.pdf

290   *Sustainability Strategic Plan 2010-2013*, Yale University, Yale Sustainability, September 2010. http://sustainability.yale.edu/sites/default/files/strategicplanupdatejune2011.pdf

291   *Our Progress: Sustainability Report 2010-2013*, Yale University, Yale Sustainability, 2013. pg 8. http://sustainability.yale.edu/sites/default/files/2013_progress_report.pdf

292   *Ibid*, pg 7.

293   *Ibid*, pg 11.

294   *Ibid*, pg 13.

NAS

Capacity Building" sets out Yale's underlying goals:

> *The success of this plan relies on system modification and behavior change. As a university with a robust culture of sustainability, Yale is able to call upon its professionals to effect change in their workplaces and in their lives while simultaneously offering students the experience of living, studying, and playing in a setting that is imbued with sustainability values.*[295]

The day after Salovey's meeting with students in the President's Room, the *Yale Daily News* reported this shift in tactics:

> *Salovey noted that while the 2010–2013 plan had focused on environmental policy changes in the Yale community, the new plan centers upon encouraging behavioral change in areas rangingfrom food consumption to paper use.*[296]

And in a letter prefacing the new Sustainability Strategic Plan, Salovey himself noted that while Yale had taken many steps to align its institutional policies with its green ideals, it had much work to do in getting its staff and students fully on board:

> *Sustainability calls for new ways of supplying energy, serving food, circulating vehicular and pedestrian traffic, distributing documents, and maintaining landscapes. We have much of the necessary technology; our challenge is to change our behaviors so that what we do with our resources provides the best stewardship for the future.*[297]

How, then, to get Yale's constituents to comply? Prod them—nudge them—towards green behavior, ideally without the students' awareness. Exactly how this might be done will depend, Salovey said, on what research in social psychology indicates will best push students' psychological buttons:

> *As a social psychologist, I am pleased that our strategies include engaging the Yale community in bringing about this change so that sustainability is embedded in the policies, practices, and day-to-day operations of our campus.*[298]

In others words, the way to green the campus is to "embed" sustainability in all aspects of day-to-day life, socially conditioning students so that they will adopt—of their own accord, or so they think— a particular habit of behavior. Sustainability becomes an active way of living to absorb, not a list of goals to examine and consider.

---

295   *Ibid*, pg 15.

296   Schwarz, "Sustainability Plan Launched."

297   *Sustainability Strategic Plan* 2013-2016, pg. 3.

298   *Ibid*.

NAS

The *Yale Daily News* reported an enthusiastic response to Salovey's plans from the Yale staff. Martha Highsmith, Salovey's senior adviser, hopes to use social media to make sustainability hip, and to rely on residential colleges to make sustainability part of the social norms of the school. She's targeting prominent members of the student body to model and to influence their peers towards lifestyleshifts.

Dining director Rafi Taherian plans to "seduce our students with plant-based foods" rather than "mandate change."[299] "This is about empowering the community to make mindful decisions and integrating these principles into everyday behavior," says Amber Garrard, the education and outreach program manager for Yale's Office of Sustainability. "Mindfulness," in this sense, refers to the actions students make, rather than the decision-making process behind those actions. Garrard wants Yalies to be mindful about recycling their paper and plastic, but unaware of the psychological manipulation going on in the background to *get* them to recycle.

> "As a social psychologist, I am pleased that our strategies include engaging the Yale community in bringing about this change so that sustainability is embedded in the policies, practices, and day-to-day operations of our campus."
>
> President Peter Salovey, Yale Sustainability Strategic Plan 2013-2014

When a campus ideology is equipped with an "education and outreach" campaign, it's clear which decision counts as "empowered" and "mindful," and which one is uninformed and inappropriate. Yale plans to immerse its students in a culture of sustainability, instructing them to play their parts until they learn them, second nature. They may have to fake it for a while, but eventually, Salovey hopes, they'll make it to full agreement with the university's principles.

Green Police

Yale's first sustainability plan, with its institutional commitments and mandates, typifies much of the first-wave environmental-sustainability movement. If the earth is burning up, its trees clear-cut, rivers drained, grounds torn, and continents overpopulated, then we must collectively halt our eco-cide or face a barren world. The kinds of actions required to avert a wholesale collapse of the ecosystem are drastic and require strict rules and enforcement.

---

299   Schwarz, "Sustainability Plan Launched."

NAS

Audi's "Green Police" advertisement during the 2010 Super Bowl satirized overzealous environmental regulations.

Audi satirized these ramrod regulations in a 2010 Super Bowl advertisement about the "Green Police," who handcuff a hapless shopper in a grocery store who falls afoul of the law by asking for a plastic bag at the checkout. A squad of police barge into a quiet home and bust the unfortunate owner of an incandescent light bulb.[300]

Two scraggy teens get caught with their illicit drinks: "What do you think of plastic bottles now?" the officer asks as he empties their disposable water bottles. But the smug owner of an Audi A3 TDI—named the 2010 "green car of the year" according to *Green Car Journal*—circumvents the long line of cars at a highway "eco-check" point. The Green Police can have no quarrel with him.

Audi's ad exaggerated but accurately identified the regulatory tactics favored by many of the environmentalist stripe. The establishment of the Environmental Protection Agency in 1970 unleashed a torrent of regulations on pollution, emissions, waste disposal, and pesticides. Some of these proved worthwhile, others overshot, but all compelled compliance.

The ban on DDT, for instance, prevented farmers from spraying their crops with the pesticide, following public outcry provoked by Rachel Carson's *Silent Spring*. Carson had argued that agricultural chemicals were destroying the environment, causing cancer in humans, and might eventually kill all the songbirds that sing each spring. The scientific evidence for Carson's claims was sparse, though.[301] The citation for the claim that pesticides obstructed human reproduction (one of Carson's major themes) was a letter to the editor of a medical journal,[302] and the claim that DDT exposure led to neurological pain in joints and limbs was based on a single complainant.[303] Carson cited Sir Macfarlane Burnet who commented in a speech at Harvard School of Public Health that increase in childhood leukemia is linked to "mutagenic

---

300   "Green Police: Audi Superbowl Ad," auto123, YouTube.   https://www.youtube.com/watch?v=Ml54UuAoLSo

301   Charles T. Rubin, *The Green Crusade: Rethinking the Roots of Environmentalism*, Lanham: Rowman & Littlefield Publishers, 1994, pp. 38-44.

302   *Ibid*, pg. 40.

303   *Ibid*, pg. 41.

NAS

stimulus," but Burnet himself dismissed pesticides and other "all-pervading elements" that suffuse the air as possible causes, because the increase was concentrated only in a subset of the populace, and was not widespread.[304] DDT had actually helped prevent rapid increase in insect populations and also limited the spread of vector-borne diseases such as malaria and yellow fever. Since the ban on DDT went into effect, an average of 2,700 daily deaths can be attributed to the *lack* of pesticides.[305]

Figure 7. Map of Key UN Sustainability Summits

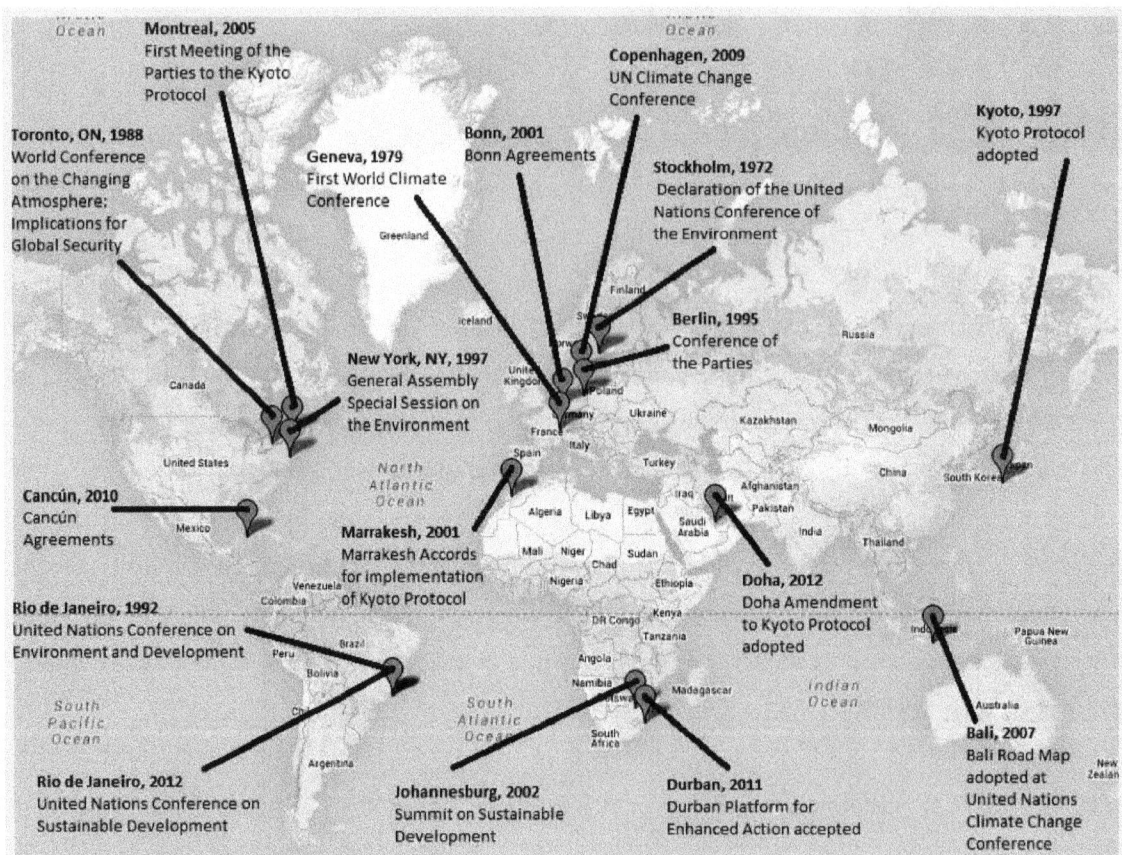

Since 1972, UN summits around the world have taken up issues of sustainability and environmental protection.

At UN summits in Rio de Janeiro (1992), Kyoto (1997), Johannesburg (2002), Copenhagen (2009), and numerous other cities, sustainability-motivated diplomats proposed mountains of new regulations demanding compliance from nations, individuals, corporations, and seemingly nature itself.

After its 2010 ad, Audi, warding off criticism from activists upset at its caricature of the green movement, was quick to note that there really are Green Police units—though perhaps not quite as zealous as their

---

304   *Ibid*, pg. 42.

305   Ed Hiserodt and Rebecca Terrell, "DDT Ban Breeds Death," *The New American*, June 6, 2013. http://www.thenewamerican.com/tech/environment/item/15583-ddt-breeds-death

Audi advertisement counterparts—in Israel, the U.K., Vietnam, and New York state.[306]

But Green Police are rare on campus. Their work is done instead by behavioral economists and social psychologists. Administrators drive students towards green behavior not by the stick, but by gentle pushes and plenty of carrots—by soft manipulation, rather than by coercion. Yale's recent shift from administrative policy to so-called "choice architecture" is a good example. The emphasis has shifted from reducing emissions and mandating student compliance to molding student sensibilities. It's a campaign that targets the subconscious.

Rhodes to Sustainability

Yale President Salovey's talk about "embedding" sustainability into the "day-to-day operations" of Yale's community is not an anomaly. Sustainability is increasingly seen as something more than a set of propositional commitments, and something closer to personal values that suffuse all areas of individual and communal life.

Frank Rhodes, former president of Cornell University, writing in the *Chronicle of Higher Education*, famously called sustainability "a new foundation for the liberal arts and sciences."[307] He characterized sustainability as a morality system "best understood within the larger framework of values, meaning, and purpose." He regretted students who "graduate untouched by the hard-won collective historical experience, social perspectives, moral considerations, and humane reflections." Sustainability, in his mind, helps to shore up students' day-to-day community as well as their understanding of human legacy and history.

John Kerry and Teresa Heinz's advocacy group, Second Nature, has similarly sought to make sustainability an inescapable part of the fabric of campus life. Second Nature aims to make sustainability "a foundation of all of an organization's activities so it becomes embedded 'in the walls.'"[308] That means, among other things, being willing to "re-imagine and reorganize the structure of the academy" so that all facets of the institution contribute to the development of a sustainable society. Teaching, research, administrative hierarchies, institutional policies, campus culture, residence life, and all areas of collegiate influence must be rebuilt and integrated together with this goal in mind.

The University of Washington presents a case study for how to embed sustainability "in the walls." In June 2014, the university was announced as one of four universities (the only one from the United

---

306   Wendy Koch, "Audi's 'Green Police' Super Bowl Ad Stirs Controversy," *USA Today*, August 8, 2010. http://content.usatoday.com/communities/greenhouse/post/2010/02/audis-green-police-ad-stirs-controversy/1#.U-DorONdX01

307   Frank H.T. Rhodes, "Sustainability: The Ultimate Liberal Art," *Chronicle of Higher Education*, October 20, 2006. https://chronicle.com/article/Sustainability-the-Ultimate/29514

308   "Institutionalizing Sustainability," *Viewpoints*, Second Nature, April 2011. http://presidentsclimatecommitment.org/files/documents/briefing-papers/institutionalizing_sustainability.pdf

NAS

States) to earn a "Sustainable Campus Excellence Award" from the International Sustainable Campus Network (ISCN), a cohort of colleges and universities that share ideas and best practices for implementing sustainability on their campuses.[309] The four annual awards recognized excellent sustainable practices in building, in campus, in student leadership, and in "integration." It was this last category of "excellence" for which the University of Washington was specifically cited.

A glance at the University of Washington's most recent Climate Action Plan provides insight as to why it excels at the "integration" of sustainability into the campus culture. Sustainability is everywhere, from increasing the cost of parking passes in order to encourage use of public transit, to recycling cooking oil as biofuel, to founding a new College of the Environment and fitting sustainability into other departments and classes that are not strictly environmental, to honoring students and staff who embody these goals with "Husky Green Awards."[310]

Indeed, the introductory note to the Climate Action Plan acknowledges that though "substantive carbon reduction" is the "primary goal" of the strategy, these carbon-reducing strategies are simply "part of a larger, more holistic set of strategies" that are aimed at

1. Moving forward toward climate neutrality

2. Engaging faculty and students in conservation and related behavior change

3. Integrating formal and informal learning on sustainability

4. Replacing the campus power plant

5. Moving students, faculty and staff to live near the UW

6. More walking/cycling, less reliance on motorized transportation

7. Becoming energy efficient[311]

Three of the seven involve changes in operations, efficiency, and energy production. But four—more than half of the primary goals of the Climate Action Plan—involve changing some aspects of students' and staff members' lives. These plans go even so far as "moving students, faculty and staff to live near the UW," presumably to cut down on emissions from commuting.

---

309 Vince Stricherz, "International Award Cites UW for Leadership in Sustainability," *University of Washington News and Information*, June 6, 2014. http://www.washington.edu/news/2014/06/06/international-award-cites-uw-for-leadership-in-sustainability/

310 Husky Green Awards 2014, University of Washington, 2014. https://green.uw.edu/hga

311 Climate Action Plan 2010 Update, University of Washington, 2010, pg. 3. http://f2.washington.edu/oess/sites/default/files/UW%20CAP%202010%20Update%20final.pdf

NAS

"Integrating" sustainability makes the university's environmental activities the personal responsibility of all faculty, staff, and students, across the board and in all aspects of campus life. Achieving this goal entails working to "Promote sustainable behavior as a cultural norm in Human Resource practices; new student orientation; faculty and staff; and in office and other work environments"—in others words, everywhere.[312] Sustainability moves beyond providing an institutional foundation; it becomes the very air its students breathe.

Commitments by Yale, the University of Washington, and others to present sustainability as a "cultural norm" indicate a new stage in the sustainability movement's progress. Early environmentalism waged war by public information campaigns and statistics, and by force: sit-ins, teach-ins, marches, and demonstrations. Think of the first Earth Day in 1970, when 20 million Americans marched down their hometown streets, occupied public buildings and held teach-ins at university lectures to grab public attention.

Early sustainability, transitioning out of environmentalism and coming out of the Brundtland Commission, progressed by edict: the international agreements and emissions caps, and other regulatory instances of the "green police." But today's sustainability is more subtle and emotional. As Yale, the University of Washington, and other colleges and universities are beginning to depict sustainability, it is a habit, a culture with its own way of living and thinking.

Bottling Sustainability

Cultural sustainability, like all cultures, has its own adages. One of them is "Think globally, act locally," a pithy imperative coined by Rene Dubos (1901-1982), the microbiologist-turned-social activist. Dubos, a French-born research doctor at Rockefeller University in New York, predates much of the sustainability movement, though sustainability mines the ore from its environmentalist predecessors as well as from its contemporary champions. Dubos meant his proverb as a criticism of globalism and a celebration of the quirks and anomalies of local life. He thought the wellbeing of the global environment came as a byproduct of local responsibility, each locale with its own customs and forms of responsible behavior.

Rene Dubos

Colleges and universities intent on sustainability have taken up Dubos's dictum. As they interpret it, the motto focuses less on local cultures and more on individual choices, regardless of their local

---

312   *Ibid*, pg. 8.

NAS

flavor. The idea is that saving the world consists in changing small, everyday habits. On one level, this teaches that the accumulated efforts of hundreds of thousands of people jointly pledging to shut off the water as they brush their teeth will, eventually, save enough water to irrigate a dry farm. On another level, it means that tinkering with seemingly innocuous, small habits will serve as the catalyst for larger lifestyle changes in the future. If the earth is to be saved, that salvation will come one fewer incandescent bulb or plastic bottle at a time.

Plastic bottles, as it happens, have been a prime target at colleges and a key force in the campaign to covertly condition students' habits and assumptions. On dozens of campuses, green-minded students and administrators denounce bottled water and shame their Dasani-toting peers into drinking from the tap. The main crime, apparently, is the bottles' wasted plastic, along with suspicions that phthalates in the plastic disrupt human hormones.[313] According to Corporate Accountability International, which operates a national "Think Outside the Bottle Campaign," 70 colleges and universities have banned outright the sale of bottled water on campus.[314] "Ban the Bottle," another anti-bottle ally, counts nearly 40 American colleges and universities as affiliates.[315] And "Take Back the Tap," a project of Food and Water Watch, counts 60 colleges and universities among its partners.[316] The tap water rebellion is vibrant on campus.

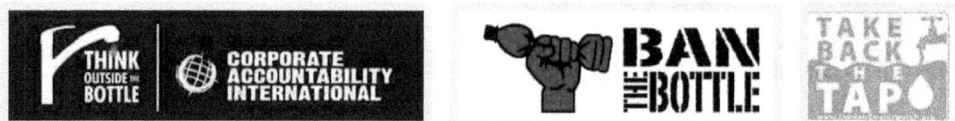

Students favor tap water to bottled water, looking to save landfill space and boycott big water corporations.

The goals of the campaign are well articulated in the short animated documentary "The Story of Bottled Water," one of several sequels to the well-known "Story of Stuff," narrated by sustainability activist filmmaker Annie Leonard.[317] In "The Story of Stuff," Leonard describes the polluted process of manufacturing the various needless products that consumer-driven Western society demands, a process created by profit-greedy corporations, she believes, and that ultimately results in needless waste, pollution, and trash. Leonard takes a similar tack in "The Story of Bottled Water." In the 1970s, she recounts, oversized capitalistic soda companies began worrying about sales declines and hit upon the idea of bottling water

313   "Bottled Water," Natural Resources Defense Council. http://www.nrdc.org/water/drinking/qbw.asp

314   "Graduating from the Bottle to the Tap," Corporate Accountability International, May 20, 2013. http://www.stopcorporateabuse. org/blog/graduating-bottle-tap

315   Ban the Bottle, Map of Campaigns. http://www.banthebottle.net/map-of-campaigns/

316   "Take Back the Tap on Your Campus," Food and Water Watch. http://www.foodandwaterwatch.org/water/take-back-the-tap/ students/

317   Annie Leonard, "The Story of Bottled Water," YouTube, 2010. https://www.youtube.com/watch?v=Se12y9hSOM0

in addition to soda. They began demonizing the tap as dirty and unsafe, and branded disposable water bottles as the safe, clean, natural way to go. The result was "manufactured demand" for bottled water as the corporations mounted a seductive, misleading advertisement campaign that persuaded people to distrust the tap and shell out their hard-earned cash for bottled versions of their own tap water.

In the eight-minute video, Leonard hits her viewers with the data (80 percent of plastic bottles end up in landfills) and attempts to chart how the bottles contribute to the degradation of the environment. But for most of the film, she avoids statistics. Facts go only so far. Leonard persuasively draws in the viewer's heart as well as his mind: "This is a story about a system in crisis," she narrates as offending disposable products zip across the screen from a dirty manufacturing center to an oversized department store, then to a suburban home crammed with stuff, and eventually to a trash heap. "We're trashing the planet" (animated trees snap off mid-trunk as a mountain peak falls off the top of the globe), "we're trashing each other" (a skull and crossbones-marked factory dumps waste into a lake where an unsuspecting gentleman stands fishing with his son) "and we're not even having fun."

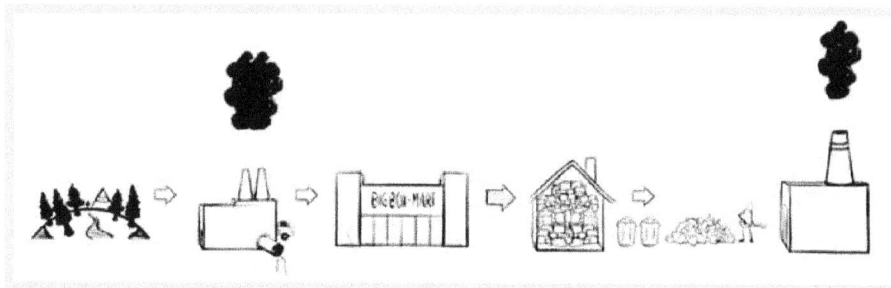

In "The Story of Bottled Water" (2010) Annie Leonard depicts a lonely materialist world of rampant consumerism.

Adopting Yale President Salovey's prodding tactic, Leonard attempts to make bottled water repulsive: "Carrying bottled water is on its way to being as cool as smoking while pregnant," Leonard says. "We know better now."

The Trayless Cafeteria

In addition to bottled water, another favorite sustainability target is the cafeteria tray, an expendable accoutrement that requires washing and enables students, who find their trays bigger than their stomachs, to take and then toss uneaten food. That waste puts the sustainability-conscientious in a dither. American University Professor Kiho Kim and AU environmental science alumnus Stevia Morawski set about testing a tray vs. no-tray cafeteria at the American University dining hall over the course of several days. Kim and Morawski measured students' leftover food and the number of dishes they used, and found that trayless dining led to a 32 percent reduction in food waste and 27 percent reduction in dish use.[318] In an article for the *Journal of Hunger & Environmental Nutrition*, they concluded that "removing trays is a simple way for universities and other dining facilities to reduce their environmental impact and save money."[319]

There's a third reason, though, besides saving money and preventing waste, that motivates institutions to sideline their cafeteria trays. The trays provide for students a strong psychological connection point between sustainability and their everyday lives. Aramark, the foodservice giant that supplies many college cafeterias, acknowledges in the opening paragraph of a 2008 report, "The Business and Cultural Acceptance Case for Trayless Dining," that

> *The increase in social consciousness and environmental stewardship on college campuses has spurred an array of new and innovative sustainability programs. One particularly creative initiative that has gained attention over the past few years is trayless dining.*[320]

Aramark identifies a number of environmental and economic reasons that might interest their clients in de-traying their cafeterias. But it also lists four reasons it categorizes under "social awareness." Trayless dining, they aver,

- Supports education and awareness of environmental issues.

- Reinforces institutions' sustainability initiatives.

- Encourages students to participate in a "green" initiative that has both a personal and community impact.

- Reinforces sustainability awareness on a daily basis.[321]

That makes the decision to serve meals on trays not just a quantitative one (how much money will we save? How much food will we save?) but a qualitative one. Using a tray or not becomes a question of values and morals, not just dollars and calories. Theo J. Kalikow, president of the University of Maine at Farmington, understood these implications when in 2007 he led his university to become one of the first to jettison its trays. "It's the right thing to do," Kalikow remarked. "Our students see sustainable practices in action on a daily basis."[322]

---

318 Charles Spencer, "Going Trayless Study Shows Student Impact," American University, January 23, 2013. http://www.american.edu/cas/news/kiho-kim-trayless-dining-hall.cfm

319 *Ibid.*

320 Aramark Higher Education, "The Business and Cultural Acceptance Case for Trayless Dining," July 2008, pg. 2. http://www.aramarkhighered.com/assets/docs/whitepapers/ARAMARK%20Trayless%20Dining%20July%202008%20FINAL.PDF

321 *Ibid*, pg. 3.

322 "College Cafeterias Ditch Trays," Natural Resources Defense Council. http://www.nrdc.org/living/schools/college-cafeterias-ditch-trays.asp

NAS

Since then, trayless dining has become only more popular. In 2009, the *New York Times* ran a front-page story on the phenomenon, commenting that "the once-ubiquitous cafeteria tray, with so many glasses of soda, juice and milk lined up across the top, could soon join the typewriter as a campus relic."[323] In 2011, the Green Report Card (a project of the Sustainable Endowment Institute that, until 2011, graded institutions on the vigor of their sustainability commitments) released its annual sustainability ranking of the 300 American colleges and universities with the largest endowments, finding that three-quarters of them had instituted some kind of trayless dining.[324] More are joining. The University of Michigan, for instance, at the start of the Fall 2013 semester scotched its trays across all cafeterias at its Ann Arbor campus, after several years of temporary trial trayless runs.[325]

The increase in trayless dining is partly due to educational campaigns, such as those conducted by American University's Kim and Morawski, that aim to demonstrate quantifiable practical benefits of de-traying a cafeteria. Four years before Kim and Morawski's study of the American University dining hall, administrators there had tried to impose a tray-ban, but the students rebelled. After Kim and Morawski showed that trayless dining did have some effect on food and water waste, students were more willing to give up their plastic trays: "This time," the university reported after Kim and Morawski's study, "without the onus of a top-down solution being imposed on them, students embraced the sustainability implications of eliminating so much waste."[326]

But more often, persuading a student to repent from his tray requires something more than cold calculations of pounds of trash averted from the landfill. It takes either a wholesale conversion of his values, or a campaign to convince him that this self-abnegation actually fits with the values he already espouses.

Yale, armed with President Salovey's social psychology research, is trying an approach that tackles both. Having been "nudged"—unaware—towards behavior changes, the student begins to reconcile his new behavior with his already-held values and, eventually, begins to assume and adopt the values behind his new behavior pattern.

---

323   Lisa W. Foderaro, "Without Cafeteria Trays, Colleges Find Savings," *New York Times*, April 28, 2009. http://www.nytimes.com/2009/04/29/nyregion/29tray.html?_r=0

324   "Food and Recycling," The College Sustainability Report Card, 2011. http://www.greenreportcard.org/report-card-2011/categories/food-recycling.html

325   Kellie Woodhouse, "University of Michigan Ditches Trays in Most Dining Halls," *Ann Arbor News*, August 26, 2013. http://www.annarbor.com/news/university-of-michigan-gets-rid-of-trays-in-dining-halls/

326   Spencer, "Going Trayless Study Shows Student Impact."

Yale first tried to go trayless in 2009, but within a week student outcry brought the trays back.[327] Two hundred students filed comment cards at the Commons Dining Hall opposing the trayless cafeteria; only six wrote in favor. Students found it difficult to carry multiple dishes in their arms. The floor grew dirty with spilled food. A group of football players piled their dishes in a tower atop a table in protest. The *Yale Daily News* reported that the Director of Residential Dining Regenia Phillips thought Yale might be able to talk the students out of their trays eventually, but only after the campus culture changed: "It won't work until it's cool not to use a tray." she said.[328]

Now, under Salovey's guidance, Yale is trying hard to make tray-ditching and other sustainability-inspired "lifestyle" changes cool. The sustainability office's assistant director Melissa Goodall even uses the word "sexy" to describe Yale's efforts to coordinate sustainability measures across the campus.[329] When Yale succeeds in making tray-ditching "cool," or "sexy," it will have fundamentally reshaped its students' values, social norms, and lifestyle habits—without lectures, data, or bothersome appeals to reason.

Mind Games

Bottled water and plastic trays are two of many minor targets that sustainability advocates are taking aim at. There are more: plastic straws, paper cups, Styrofoam to-go boxes, envelopes, plastic grocery bags, to name a few.

These seemingly trifling measures would not halt man-made global warming. American University estimates that it might save 25,000 pounds of food scraps per year by purging trays,[330] and U.S. water bottle consumption might add up to millions of bottles of a year.[331] Those are big numbers. But these token contributions are minuscule relative to the vast gulf sustainability advocates see between where society is and where a green, no-footprint society ought to be. An "impact-neutral," "no-footprint" society would require giving up cars, refrigerators, airplanes, and many digital devices—not just the lower-threat bottled water or plastic trays.

Consider the trays. The amount of effort required to give up a tray might have been more efficiently invested in, say, biking to work once a week, or simply replacing an outdated, inefficient heating unit.

The cafeteria trays' contribution towards the greening of the country is infinitesimally small relative to the inconvenience of abiding by this new eco-morality. That imbalance leads one to wonder whether it's not

327 Nora Caplan-Bricker, "Return of the Trays," *Yale Daily News*, September 4, 2009. http://yaledailynews.com/blog/2009/09/04/return-of-the-trays/

328 *Ibid.*

329 Schwarz, "Sustainability Plan Launched."

330 Spencer, "Going Trayless Study Shows Student Impact."

331 Annie Leonard, Story of Stuff Project, Youtube, 2010. https://www.youtube.com/watch?v=Se12y9hSOM0

NAS

the environment but the inconvenience itself that is the primary goal of these campus exercises in self-deprivation.

The student who navigates the cafeteria with his arms full of plates, cups, and cutlery, dripping soup and spilling soda, is not saving much dishwashing water, or even all that much food, compared to the absolute quantities of dining hall food purchases. His trayless juggle is not necessarily making him any healthier, nor is it doing much to help him combat the freshman-fifteen, as some sustainability activists have suggested. (In fact, a study from two Cornell economists found that students without trays tend to run out of hands and to skip extra dishes—usually healthy dishes such as salads—in order to better carry their entrée and dessert. This leads to students consuming relatively fewer greens and more sweets.[332])

> *Students are taking small measures with upfront inconveniences that habituate them into a larger lifestyle replete with sustainability-minded activity.*

But the student in the midst of that juggling act is accomplishing one crucially important objective. He is aware three times every day that he is making a sacrifice on behalf of the environment. He is taking small measures with upfront inconveniences that jar him alert to the need for other, larger measures he may take in the future. He is being habituated into a larger lifestyle replete with sustainability-minded activity. Today he'll eschew the tray and sip a glass of water from the tap; tomorrow he'll bike to campus; and in thirty years, he'll commit his hedge fund to invest in alternative energy.

Long-term behavior modification is exactly what Yale and its compatriot schools have in mind. The text of Yale's Strategic Plan is itself quite clear in its language. Some of the intermediate goals include:

> *Establish a culture of green information technology at Yale through a portfolio of training and certification programs, as well as consistent online messaging to all computer-using members of the Yale community by June 2016.*[333]

Or

> *Promote sustainability as a core business value at Yale by June 2016.*[334]

---

332   Brian Wansink and David R. Just, "Trayless Cafeterias Lead Diners to Take Less Salad and Relatively More Dessert," *Public Health Nutrition*, November 2013. http://journals.cambridge.org/action/displayAbstract?fromPage=online&aid=9073711&fileId=S1368980013003066

333   *Sustainability Strategic Plan 2013-2016*, pg. 16.

334   *Ibid*, pg. 15.

NAS

Presumably a "culture of green information technology," "consistent online messaging," and the sacralizing of "sustainability as a core business value" will cultivate in students a form of thought and of life that remain with them long after they leave Yale's campus.

The University of Texas at Arlington has something similar in mind. UT Sustainability Director Meghna Tare puts the objective more directly on the sustainable business blog "Triple Pundit":

> Students attending a university that places high value on sustainable operations are more likely to take this mindset to their future places of employment where they can help shape the future of environmentally-friendly companies.[335]

Tare's office at UT Arlington happily "embraces" its role of teaching students "the knowledge, skills, and habits to help society shift to a more sustainable world, both in their professional and personal lives" (emphasis added). To help engrain those habits in students' lives, the Office of Sustainability holds energy saving competitions to entice students with green living, sponsors student clubs, holds attention-grabbing campus events, and co-opts the curriculum to surround students with an environment rich in the rhetoric of sustainability.

The Competitive Drive

Demanding sacrifices of simple conveniences such as cafeteria trays and water bottles can wear a student down. Fortunately, sustainability offers incentives as well.

In January 2001, two college campus recycling coordinators found themselves at a loss for how to get their students to care—let alone get excited about—recycling their trash. Working memory can only hold so many thoughts at once, and the act of recycling—having to consciously consider their trash, separate it from recyclables, recall which items could be recycled, and seek out a recycling bin—add up to a series of steps that, if not routine by force of habit, requires just enough effort to become a pesky inconvenience.

The recycling coordinators, one from Ohio University, the other from Miami University, hit upon the idea of making recycling a competition between the two schools. Ohio and Miami have a longstanding sports rivalry; beating the other school was a source of pride and school spirit. The recycling competition tapped that school spirit, turning sustainability into a source of institutional honor and inclusion with the campus community. And because the terms of the competition awarded the win to whichever university scored the highest rate of recycling per person, rather than in absolute numbers, recycling became a personal duty to preserve the school's reputation.

---

335   Meghna Tare, "Fostering Sustainability in Higher Education: University of Texas at Arlington," *Triple Pundit*, December 3, 2013. http://www.triplepundit.com/2013/12/fostering-sustainability-higher-education-university-texas-arlington/

NAS

Over the course of ten weeks, university staff monitored recycling bins and heavily promoted recycling to the student body. Ohio recycled an average of 32.6 pounds per person; Miami, 41.2 pounds—and with it gained the recycling victory.[336]

RecycleMania has since grown to a nationwide annual tournament sponsored in part by the EPA WasteWise program and operated by the national nonprofit Keep America Beautiful. In 2014, a total of 461 American and Canadian colleges and universities clashed in trash warfare. The tournament operates on a rulebook for measuring and reporting recycling rates. It offers 11 categories of competition from specific targets (paper, cardboard, cans, etc.), to the Gorilla prize awarded for highest gross tons of recycled material, to the Grand Champion, which denotes the university with the highest ratio of recycled waste to trash. For broader reach, RecycleMania offers two divisions: "competitive" for those seeking to win and willing to carry out the detailed measuring rules, and "benchmark" for those who will report less precise data, forgo the competition, and simply want to encourage their students to recycle.

The event proves successful. The 2014 Grand Champion, Antioch University in Seattle, managed to recycle 93.133 percent of all waste. The University of Missouri-Kansas City took second with 81.052 percent of all trash recycled during the competition period.

Antioch broke the 90 percent mark with the help of a few heavy-handed tactics: "we made it harder to throw things out by actually removing nearly all of our trash cans," Antioch President Brian Baird commented in an announcement of the win.[337] But its goal superseded a RecycleMania win: "Even though we are a socially conscious institution that lives by our mission, we made people think even harder than they normally do, and make decisions item by item." Jennifer Jehn, president and CEO of Keep America Beautiful, echoes Baird's sentiment: "Recyclemania is a powerful tool to communicate the recycling message to college students in a way that resonates with their values and experience."[338]

---

336   History of RecycleMania. http://recyclemaniacs.org/about/history-recyclemania

337   "Waste Not! Antioch University Seattle is Tops in Recyclemania," Antioch University Seattle, March 18, 2014. http://www.antiochseattle.edu/2014/03/2014-recyclemania-happening-now-2/

338   *Ibid*

RecycleMania is one of hundreds of competitions held on campuses every year. There are dorm competitions to reduce electricity (such as the Kill-a-Watt at the University of Michigan), campus events to promote recycling awareness (trash fashion shows, where students model outfits constructed from newspapers, lamp shades, and other items destined for landfills), and scavenger hunts that advertise campus solar panels and bike repair shops. If cafeteria trays and water bottles are the gentle sticks that prod students toward sustainability, competitions provide the prizes that draw them forward.

"Eco-Reps"

The social architect cannot quite create social norms at the snap of his fingers, though tinkering with authority structure—or cafeteria equipment—can do quite a lot. Creating a fully immersive sustainability experience requires stocking a campus with at least a few true believers ready for the task of quiet evangelism.

On many campuses, this position is called an "eco-rep" or an "environmental ambassador." Sometimes the titles are rather militant, as at Yale, where these students form the "Sustainability Service Corps." Other times the titles take on a clinical feel. At Indiana University, the Office of Sustainability hires six students for ten hours a week at a rate of $10 per hour as part of its bluntly-put "peer educator program."[339]

"Eco-reps" are students, generally paid by their university's Office of Sustainability, who have bought the sustainability dogma wholesale and fully internalized it. They advise their peers about recycling, shorter showers, Meatless Mondays, and other ways to grow greener. The Association for the Advancement of Sustainability in Higher Educationcountseightyinstitutionsthathave student environmental representatives of some kind.[340]

Eco-Reps monitor their campuses at the University of Massachusetts-Amherst, Bowdoin College, Cornell University, Clemson University, the University of St. Louis, and elsewhere.

---

339   Sustainability Peer Educator Program, University of Indiana. http://hoosierbiology.wordpress.com/2014/01/06/sustainability-peer-educator-program/

340   "Student Peer-to-Peer Sustainability Education Programs," Association for the Advancement of Sustainability in Higher Education. http://www.aashe.org/resources/peer-peer-sustainability-outreach-campaigns

Northwestern University has 60 such "Eco-Reps" who focus on "empowering students that aren't already engaged in the environmental movement, making sure they have the necessary resources to make greener choices."[341] During the 2013-2014 school year, they promoted recycling, talked up sustainability among their peers, and held competitions and events to spark students' interest.

Northwestern Eco-Reps also held a "Living Green Fair" to spotlight all the ways to live an eco-friendly life and ask their peers to sign a green pledge in which they select various ways to conserve energy, reduce waste, eat sustainably, save water, and travel sustainably. The preface to the pledge lays out the importance of personal behavior:

> *Northwestern is committed to being a leader in sustainability – in our operations, research, and curriculum as well as in the community – and we're asking you to commit to simple choices that together, will help us achieve this goal and make Northwestern a healthier, moresustainable environment and a better globalcommunity.*[342]

With friends and peers signing pledges, and with the reputation of their university's status as a "leader in sustainability" on the line, who but the most staunchly opposed could resist?

Harvard's Eco-Reps (or Eco-REP, for Resource Efficiency Program, as the sustainability ambassadors are called there) have a particular claim to effective cultural change. In an article titled "University Looks to Students in Effort to Drive Down Waste," the *Crimson* reported that Harvard's waste reduction and recycling rates had recently improved. The *Crimson* reported on the Eco-REPs' success in this happy trend:

> *Jaclyn Olsen, assistant director of the Office for Sustainability, attributed improvement in recycling to similar efforts, but said that a change in culture awareness has also played a part. "I really think it's been a big culture change," Olsen said of students who have grown more environmentally conscious.*[343]

The REPs set about making their peers more aware of recycling, modeling responsible trash habits, campaigning against bottled water, and trying to make recycling more convenient. They reached out to student groups already active on environmental issues, and together worked with the administration to develop new ways to promote recycling. And they realized how cautious they had to be when pressuring their peers towards a new habit. "If you are trying to change someone's lifestyle...people are furious

---

341 "Eco-Reps Lead the Charge Towards a Greener NU," Northwestern University Office of Sustainability. http://www.northwestern.edu/sustainability/news/2014/articles/eco-reps-lead-the-charge-towards-a-greener-nu.html

342 "Sustain NU," Green Pledge, Northwestern University Office of Sustainability. http://greenpledge.fm.northwestern.edu/

343 Noah J. Delwiche, "University Looks to Students in Effort to Drive Down Waste," *Harvard Crimson*, March 27, 2014. http://www.thecrimson.com/article/2014/3/27/harvard-sustainability-groups/?page=single

NAS

about that," said Kristen J. Wraith, a three-year veteran Eco-REP and former chair of the Environmental Action Committee.[344]

Gambling with the Future

In some ways, it's a gamble that Harvard, Yale, and their counterparts take. There is no guarantee that the habits cultivated by doffing the tray or winning a recycling competition will stick, or that if they do those habits will swell into larger lifestyle changes. One sustainability director tried a similar social psychological nudge approach to her students and found herself fired after a year and a half when the experiment failed to achieve measurable results.[345]

Towson State University released sustainability director Clara Fang in October 2013 for what she described as too much emphasis on student involvement and too little on institutional changes. Fang wrote on the "Towson Goes Green" blog that, "according to my supervisors, I was not fulfilling my duties as a facilities management employee due to a focus on student outreach."[346] Fang had prepared a Climate Action Plan to achieve climate neutrality in 2050, a greenhouse gas inventory in 2010 and 2011, and a Waste Reduction strategic plan, but evidently spent the bulk of her time on initiatives meant to attract students: recycling competitions, student sustainability events and conferences, a residential recycling initiative, an eco-reps program, a sustainability interns program, and other projects meant to draw students' interest. These hadn't taken root fast enough to lead to substantial measurable differences.

For that reason, the slow, psychological prodding approach has faced criticism from some sustainability advocates. Adam Corner, a psychology researcher and sustainability supporter, wrote an article on *The Guardian*'s blog "Sustainable Living Hub," that declared, "'Every Little Helps' Is a Dangerous Mantra for Climate Change."[347] Telling people to take small steps towards saving the earth might send the wrong psychological messages, quieting their consciences with the salve of penance rather than spurring them with the weight of ecological guilt. Corner considers the popular sustainability target of plastic grocery bags, though he might as easily have looked at café trays:

> Like recycling, re-using carrier bags has become something of an iconic "sustainable behaviour". But whatever else its benefits may be, it is not, in itself, an especially good way of cutting carbon. Like all simple and painless behavioural changes, its value hangs on whether it acts as a catalyst for other, more impactful, activities or support for political changes.

---

344   Ibid

345   Jonathan Munshaw, "Sustainability Says Bye to Second Director in Three Years," *The Tower Light*, October 23, 2013. http://www.thetowerlight.com/2013/10/sustainability-says-bye-to-second-director-in-three-years/

346   Clara Changxin Fang, "A Message from TU's Departing Sustainability Manager," Towson Goes Green, October 21, 2013. http://greentowson.wordpress.com/

347   Adam Corner, "'Every Little Helps' Is a Dangerous Mantra for Climate Change," *The Guardian* Sustainable Living Hub, December 13, 2013. http://www.theguardian.com/sustainable-business/plastic-bags-climate-change-every-little-helps

NAS

Whether reusing plastic bags becomes a habit that leads to "other, more impactful, activities" or "political changes," depends on the person's psychological makeup and how deeply his current habits and belief structures are rooted. If the habit stops at reusing plastic bags, and doesn't develop into an aversion for plastic, a willingness to take public transit, votes for the Green Party, and installation of home solarpanels, then the drive to reuse plastic bags becomes an inconsequential accident, rather than the source of a central worldview. "Nudging, tweaking, or cajoling people into piecemeal behavioural changes like re-using plastic bags is not a proportionate response to climate change," Corner writes. Instead,

> *Engaging the public through their personal carbon footprints is really only a means to an end – and that end is a political and economic system that has sustainability as its central organising principle. And if these sound like radical statements, unbecoming of the stately, reserved sentiments associated with the Royal Society, then consider the prospect of a world that is four or even six degrees hotter and the havoc and suffering that would be inevitable. This is also a radical choice.*

Moral Habituation

Cultivating dispositions and instilling moral principles by habit and practice is not foreign to education, of course. It is a very old practice, one rooted from antiquity at the very heart of education itself. Aristotle advocated a kind of personality sculpting and conduct formation in the Nicomachean Ethics: "Virtue of character results from habit....A state of character arises from the repetition of similar activities." There are religious grounds for cultivating habits, as well. In Judaism, orthodoxy involves a great deal of orthopraxy: revering the Torah and Talmud involves, primarily, obeying them. In Christian teaching, the spiritual regeneration that so confused Nicodemus initiates not only atonement between God and man, but also the conversion of one's soul and the conduct-shaping principles that animate it.

> *"To my mind the real greatness of a nation, its true civilization, is measured by the extent of this land of Obedience to the Unenforceable."*
>
> *Lord Moulton, "Law and Manners"*

Between the realms of positive law and license lies the intermediate domain of habit, governed, as Lord Moulton said, by "obedience to the unenforceable." Here no written or enforced rules apply, yet other unwritten rules guide one's behavior. Manners, civic duties, and ethics weigh on one's conscience. The lack of police force is key: "To my mind the real greatness of a nation, its true civilization, is measured by the extent of this land of Obedience to the Unenforceable," Moulton remarked in a speech, "Law and Manners," delivered before the Authors Club in London around the turn of the twentieth century. "The

NAS

true test is the extent to which the individuals composing the nation can be trusted to obey self-imposed law."[348]

Yale and its sustainability counterparts see their project as one that encourages the inclusion of certain "unenforceables" in campus norms. Elevating trayless dining to the realm of law backfired at Yale four years ago, so for now,   the decision to forgo trays remains in the realm of free choice. But manipulate the social atmosphere and—*voila!*—a new social norm appears, fed and supported by peer pressure as students willingly impose on themselves the strictures they once bucked.

The campus sustainability movement's stratagem effectively abolishes some measure of   student freedom: technically Yale undergraduates could undermine their administrators' systems by carrying trays and stopping their ears during sustainability orientation lectures, but social norms effectively hedge them in. More concerning is the subterfuge eating away not at choice per se, but at conscious, rational   choice.

Yale's strategy signals an expansion of the domain of unenforced law—something Moulton saw as a sign of true civilization in the aggregate and trustworthiness in the individual. But where Moulton characterized these morals as primarily fixed standards of human decency, Yale's sustainability plan draws on a partisan ideology: sustainability. Sustainability's vendetta against capitalism and against social rigidity engenders its own lifestyle and habits, to be sure, but these hardly parallel the timeless traits of a virtuous, classical gentleman.

A linguistic clarification helps. "Mores" (from the Latin for manners) refers to conventions familiar and customary to a particular culture. Ethics comprises cross-cultural absolutes. The former indicates commonplaces like shaking hands and tipping your hat—or recycling your plasticware and eating vegan. The latter declares that murder, stealing, and idolatry, to name a few, are wrong. Where mores match ethics, social conventions encourage (quite helpfully) the simultaneous formation of good citizens and good men. Where they do not, men risk obedience to one at the expense of the other. "What ally should I invoke," Antigone cries, having buried her brother in defiance of Theban King Creon, "when by piety I have earned the name of impious?"

Giving up a lunch tray isn't an ethical duty—and as of now, it isn't a moral duty either. Trayless dining hasn't quite reached the realm of normal, basic assumptions; it's still new and surprising enough that the *New York Times* could print a front-page article on the trend. But if Yale and its counterparts succeed, it might at some point become the norm.

---

348   Lord Moulton, "Law and Manners," *Atlantic Monthly*, July 1924.

Manipulation

Therein lies the fault. Yale—and its counterparts—is artificially creating social norms meant to architect student's choices, tamper with their lifestyles, bombard them with promotional material, and nudge them towards the desired response. The movement does not initiate—with conscious ceremony or formal statements to affirm—its novices into its ranks. It conditions them, subtly, into a service kept secret from even their own sensibility.

Manipulating a pupil into "good" behavior seems hypocritical at best, nefarious at worst. But where virtue ethicists see ethics as a set of ultimate, often sacred precepts, recent social science forays into the realm of human habituation—such as those Salovey is trying out on the Yale campus—focus on reflexive, subconscious reactions to social stimuli. In this line of thinking, the primary forces that shapeour habits and our characters are contingent happenstances erected by social hierarchies. These hierarchies constrain us more often than they liberate us, and because they are artificial and secular, rather than rational and sacred, nothing prevents our tampering with them. If our characters and habits result from social structures, themselves the malleable products of our own actions, then they can and perhaps even ought to be tempered and re-forged according to a more progressive template. Utility-motivated manipulation (the kind that Yale is advocating) becomes acceptable—at least to the manipulators.

So how do colleges and universities construct this atmosphere suffused with sustainability that nudges their students towards green living? Salovey cited social psychology research broadly speaking. He might have cited more specifically *Nudge*.

A 2008 *New York Times* bestseller, *Nudge* could serve as the playbook for the sustainability movement's recent advances on college campuses. *Nudge* advises strategies to tweak our habituating structures to push us towards the best choices. In it, Richard Thaler (professor of behavioral sciences and economics at the University of Chicago) and Cass Sunstein (a Harvard law professor who then became administrator of the White House Office of Information and Regulatory Affairs under President Obama) distinguish between "Econs" and "Humans."

"Econs" are the fully rational, computing robots that economists assume humans to be; "Humans" are the emotional, contextual creatures we really are. We Humans act on impulse, use rules of thumb, inertly favor the status quo, and neglect to study all of our

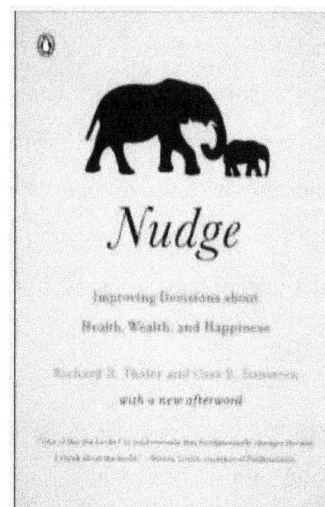

options, and so we often choose poorly. Educating Humans to be Econs isn't feasible, but teaching them to mimic Econs is. If not all of us can decide in a rational manner, then the rational ones among us may as well nudge the others towards Econ-certified rational behavior.

Thaler and Sunstein advocate for a "libertarian paternalism": libertarian because of "the straightforward insistence that, in general, people should be free to do what they like" and paternalistic because "it is legitimate for choice architects to try to influence people's behavior in order to make their lives longer, healthier, and better."[349] These paternalistic "nudges" leave people free to choose among a preset buffet of options within an intentionally architected environment. The social planner arranges the circumstances of people's choices—making desserts harder to find in the cafeteria, or publishing a list of the highest polluting domestic manufacturers—in order to encourage the socially optimal choice. "To count as a mere nudge," Thaler and Sunstein assure their readers, "the intervention must be easy and cheap to avoid. Nudges are not mandates."

"Easy and cheap to avoid"—if you're an Econ, that is. If, as Thaler and Sunstein establish early in their book, Humans gravitate towards the default on everything from magazine subscriptions to health insurance policies, then evading nudges isn't so simple. Nudging Human behavior cracks open the door—if not entirely unhinging it—to the danger of authoritarian abuse. And even if the social architect nudges selectively, creating only those mores that match ethics, the nudged respondents may act, but not be, virtuous. Virtue requires virtuous intentionality, not just one-off activities.

The sustainability movement's engineering of the social environment to protect the natural one treats human choice as a programmable response to outside stimuli. There's nothing virtuous, except in a utilitarian weighing of outcomes, about automatically reacting to external cues.

The college campus sustainability commitment represents a significant shift in higher education, away from educating students with rational and moral knowledge that prepares them for wise, conscious choices, and towards covert training operations that elicit Pavlovian responses. The mark of good character used to be one's trustworthiness, as Moulton put it, in choosing the right behavior of one's own accord, no nudging required. Now, apparently, it's sufficient to play the part.

If Yale were nudging its students towards politically incorrect mores—say, encouraging female students to join the cheer squad and avoid sports, or shaming international students who eat their preferred ethnic food—there'd be an outcry: Interference with lifestyle! Intolerance! Regressive social norms! But when a

349   Thaler and Sunstein, *Nudge*.

NAS

politically correct dogma is forced on unsuspecting students, no one calls foul. Perhaps any would-be discontents are too busy faking it.

CHAPTER 5: AMAZING WASTE: WHAT SUSTAINABILITY COSTS

Towards the middle of campus, south and a little to the east of the inner ring of buildings that surround the Middlebury College quad, an angular building rises from the earth. Through its glass plate casing passersby can see white utility pipes running from the walls and ceiling into a gigantic blue cylinder flanked by red utility boxes and surrounded by grey walkways and stairwells. The industrial building contrasts sharply with the homely piety of the neighboring cemetery, where many of the college's founders are buried (along with the cremated ashes of one unfortunate Egyptian mummy that fared poorly in New England's humidity), and with the stately dignity of the marble, neo-colonial McCullough Student Center across the street.

This steel and glass building houses Middlebury's new Biomass Gasification Plant, a wood-fed boiler engine that produces steam to heat and power much of Middlebury's campus. Inside the glass walls, wood chips chopped to no more than two inches thick and dried to 50 percent moisture content ride a conveyer belt into the cylindrical boiler, where they smolder under extreme temperatures and emit wood gas that, when mixed with oxygen, ignites to 1100 degrees Fahrenheit.

Middlebury College's $12 million biomass gasification plant moves the college closer to carbon neutrality.

Metal tubes carrying water crisscross the interior of the cylinder, and as the water inside turns to steam, the energy—30 million BTUs per hour, 15-20 percent of the college's total electricity needs—powers campus heating units, hot water heaters, and other electrical equipment to keep Middlebury's campus warm and fed with electricity.[350] After dark, even on the most frigid days of Vermont winters, the window panes, like a nightlight, glow with activity, casting comforting fingers of light across the parking lot and the Old Chapel Road behind it.

The Gasification Plant serves as a sign of hope to Middlebury residents in another way, too. It is carbon

---

350   "Middlebury College's Biomass Heating and Cooling Plant Aims to Cut Carbon and Costs—in Big Ways," Biomass Energy Resource Center, Biomass Case Studies Series, 2008. http://www.biomasscenter.org/images/stories/middlebury-college.pdf

NAS

neutral, a key achievement on Middlebury's path to eradicating or offsetting 100 percent of all campus carbon emissions by the year 2016. The gasification plant relies on trees (theoretically, constantly replanted and therefore renewable) that are harvested and shipped—20,000 tons every year, up to two or three truckloads a day—to Middlebury from within a 75-mile radius. Trees absorb carbon from the air and store it inside their trunks, and when they die and decompose—whether by decay on a forest floor or combustion in an engine room—they release the carbon back into the atmosphere. The gasification plant releases no more carbon than the trees would have emitted had they rotted in the forest, and so long as the College replants what it cuts down, the local carbon levels should, in theory, remain stable. And because the wood chips smolder in low-oxygen tanks, rather than burning as in an old-fashioned stove, no smoke emanates from the plant. Middlebury filters its exhaust assiduously, with 99.7 percent accuracy in removing all particulates, so the gases pouring out of the stacks into the sky are almost entirely steam.

Going zero-carbon means a chance at admittance to an exclusive group of environmental enthusiasts whose membership, so far, includes only three: tiny, 300-student College of the Atlantic in Maine (2007),[351] Vermont's Green Mountain College (2011),[352] and Colby College, also in Maine (2013).[353] One other university had previously announced its carbon neutrality, but later had to retract that claim when it came to light that both the university and the local electricity provider had inadvertently been double-claiming the same renewable energy credits. Winning that fourth spot is a big race between the Presidents' Climate Commitment signatories, which has now grown in number to 685. All of them have pledged to eliminate their net greenhouse gas (GHG) emissions entirely, by "minimizing GHG as much as possible, and using carbon offsets or other measures to mitigate the remaining emissions."[354]

Middlebury's pledge—carbon neutral by 2016—is ambitious. Our environment is built on carbon, and the economy runs on fossil fuels. Renewable energy remains costlier than oil and gas. Targeting zero-carbon requires changing staff and student habits and instituting new training. Infusing sustainability throughout the campus culture means establishing an office of sustainability, hiring new staff, adding new degree programs and incentivizing professors to re-focus their courses. Just how costly sustainability initiatives are, though, is not often transparently analyzed within the field.

Footprints in the Earth

"Carbon neutrality" sounds benign and simple enough, as if it required nothing more than to stop doing

---

351   Kenny Luna, "And the First Carbon-Neutral College Campus in the US Is…," *Treehugger*, December 21, 2007. http://www. treehugger.com/corporate-responsibility/and-the-first-carbon-neutral-college-campus-in-the-us-is.html

352   "Climate Neutrality," Sustainability at Green Mountain College. http://sustainability.greenmtn.edu/climate_neutrality.aspx

353   Ruth Jacobs, "Colby Achieves Environmental Milestone: Carbon Neutrality," Colby College News, April 4, 2013. http://www. colby.edu/news/2013/04/04/colby-achieves-environmental-milestone-carbon-neutrality/

354   Frequently Asked Questions, American College and University Presidents' Climate Commitment. http://www. presidentsclimatecommitment.org/about/faqs#10

something unnatural, or to cease swinging between extremes and to settle peaceably at an equilibrium. In fact, it's something of an economic black hole. Full elimination of carbon emissions requires an expensive overhaul of campus life. Institutions must retrofit or even reconstruct campus buildings to reflect cutting-edge efficiency technologies, and replace college vehicles with electric or hybrid biodiesel cars. They have to figure out how to cut down on administrative air travel or else purchase carbon credits to offset the emissions.

Even simple tasks such as ordering paperclips and sticky notes for the admissions office become fraught with ethically-charged factual questions: *Was the iron mined without destroying a mountainous ecosystem? Did the paper come from "sustainably" managed trees? How far did the delivery truck travel, and is there any way to order from someplace closer?* Middlebury's "Procurement Policies and Procedures" manual, for instance, authorizes college representatives to give preference to local, minority-owned businesses that "demonstrate superior long-term sustainability, energy efficiency, and pollution minimization in product production and usage life cycles."[355] Middlebury's "Recycled Paper & Purchasing Policy," adopted in 2007, specifies minimum percentages of post-consumer waste recycled product in its paper (100 percent for all College office uses), asks all employees to consider printing fewer pages, and requires that the "stock will be readily and consistently available from a local supplier."[356]

The EPA's "Environmentally Preferable Purchasing" guidelines, used by many universities[357] as a supplement to their own purchasing guidelines, go further. The guidelines recommend that solid Polyethylene (PE) plastic binders, for instance, should have at least 30-50 percent post-consumer recycled material in them, and at least 30-50 percent of the material should be recoverable for recycling again.[358] But solid Polyethylene Terephthalate (PET) should be made of 100 percent recycled material and be 100 percent recoverable, while paper-covered binders should hold 75-100 percent recycled material and be 90-100 percent recoverable. The guidelines come with detailed charts breaking down precise standards for all office supplies.

355 "Procurement Policies and Procedures at Middlebury College," Middlebury College, Vermont, pg. 11. https://www.middlebury.edu/media/view/252745/original/ProcurementPolicy.pdf

356 "Recycled Paper and Purchasing Policy," Middlebury College, 2007. http://www.middlebury.edu/sustainability/policy-planning/policies/paper

357 For instance, "Duke University Stores Green Purchasing Policy," Duke University Office of Sustainability. https://sustainability.duke.edu/documents/Duke%20Stores%20Purchasing%20Policy.pdf

358 "Non-Paper Office Products," Environmentally Preferable Purchasing, Environmental Protection Agency. http://www.epa.gov/epawaste/conserve/tools/cpg/products/nonpaperoffice.htm

NAS

Figure 8. EPA Purchasing Guidelines[359]

EPA's Recommended Recovered Materials Content Levels for Binders, Clipboards, File Folders, Clip Portfolios, and Presentation Folders			
Product	Material	Postconsumer Content (%)	Recovered Materials (%)
Binders - Plastic Covered	Plastic	- -	25–50
Binders - Paper Covered	Paper	75–100	90–100
	Pressboard	20	50
	HDPE	90	90
Binders - Solid plastic	PE	30–50	30–50
	PET	100	100
	Misc. plastics	80	80
	HDPE	90	90
Plastic clipboards	PS	50	50
	Misc. plastics	15	15–80
Plastic file folders	HDPE	90	90
Plastic clip portfolios	HDPE	90	90
Plastic presentation folders	HDPE	90	90

EPA's Recommended Recovered Materials Content Levels for Office Recylcing Containers and Office Waste Receptacles			
Product	Material	Postconsumer Content (%)	Recovered Materials (%)
	Plastic	20–100	- -
	Steel[1]	16	25–30
Waste Receptacles	Paper:		
	- Corrugated	25–50	25–50
	- Solid Fiber Boxes	40	- -
	- Industrial Paperboard	40–80	100

359 Ibid.

NAS

Recommended Recovered Materials Content Levels			
Product	Material	Postconsumer Content (%)	Total Recovered Materials Content (%)
Furniture structure	Steel[1]	16	25–30
Furniture structure	Aluminum	- -	75–100
Cellulose Loose-Fill and Spray-On	Postconsumer Paper	75	75
Particleboard/Fiberboard component[2]	Wood or wood composite Agricultural fiber	Greater than 0 - -	80–100 100
Fabric	PET	100	100
Plastic furniture component	HDPE	70–75	95
Remanufactured or Refurbished Furniture	Various	25–75	25–75

Carbon neutrality quickly becomes invasive, too. In calculating an institution's net greenhouse gas emissions, the American College and University Presidents' Climate Commitment includes student, staff, and faculty commutes, obliging signatory schools either to mandate green transportation or to purchase carbon credits that off-set their students' and staff's behavior.[360] In that spirit, Middlebury asks its students to consider purchasing carbon offsets to make up for their travel when they study abroad,[361] and asks incoming students to buy $36 in carbon offsets to counteract the 3 tons of carbon that an average student will consume by living in a campus dormitory.[362]

Big Footprints

In hoping to erase carbon footprints, sustainability leaves an economic footprint of its own. Going green is pricey. Exactly how pricey, though, is hard to find out. Those economic footprints are kept carefully hidden. When one goes looking for realistic and reliable estimates of what sustainability efforts cost colleges and universities, there are few ready-made answers to be found.

This is ironic in that the movement frequently criticizes free market capitalism for its failure to account

360   The ACUPCC asks institutions to include almost everything in their assessments: "At a minimum, participating campuses should include in their inventories: (1) direct emissions produced through campus activities (known as "Scope 1 emissions"); (2) indirect emissions from purchased energy ("Scope 2"); and (3) indirect emissions from (a) student, faculty, and staff commuting; and (b) institution-funded air travel ("Scope 3"). As the inventory methodology develops and to the extent practical, participating institutions should also endeavor to evaluate embodied emissions in purchased goods and services, including food." "What emissions sources are included, and how are they calculated?" FAQs, American College and University Presidents' Climate Commitment. http://www.presidentsclimatecommitment.org/about/faqs#11

361   "Carbon Offsets," Middlebury College, Study Abroad. http://www.middlebury.edu/international/sa/sustainable/carbon_offsets

362   Tess Russell, "Offsets Figure into Carbon Neutrality Plan," *The Middlebury Campus*, February 14, 2008. http://middleburycampus.com/article/offsets-figure-into-carbon-neutrality-plan/

NAS

for the total life-cycle costs of products "from cradle to grave." A favorite talking point of sustainability advocates is that the environmental costs of extracting raw materials and disposing (or recycling) items that are no longer useful isn't adequately reflected in market-based prices. Yet when it comes to the cost of the sustainability movement itself, this love of transparency simply disappears. No one knows exactly what sustainability costs—except that the figure is very high.

It is telling that the Presidents' Climate Commitment requires its signatories to make numerous environmental impact statements available for public review and verification, but none on financial viability. Each signatory completes an initial "Implementation Profile" summarizing what tactics of greenhouse gas (GHG) elimination the school plans to use, a "Climate Action Plan" detailing these actions, periodic "GHG reports" chronicling the campus's environmental improvement (usually accompanied by a massive thousand-cell "Inventory Calculator" to gauge impact and a supplemental "Inventory Narrative" kindly interpreting the data), and finally a "Progress Report" to summarize all consequent environmental enhancement, student and faculty activism, and community outreach. Of these six documents, only the Progress Report asks schools to estimate the money spent and saved through sustainability measures. That Progress Report has five sections. One-half of one section, or one-tenth of the report, deals with finances, and the questions there are optional.

The ACUPCC on its website prominently promises signers that "exerting leadership in addressing climate disruption" will not only fulfill "an integral part of the mission of higher education" and "attract excellent students and faculty," but also "stabilize and reduce their long-term energy costs" and "attract new sources of funding."[363]

Whether sustainability programs reduce operating costs remains to be seen. Some high-profile institutions, however, have indeed secured outside funding. In May 2014, California Technical Institute brought in $15 million from Lynda and Stewart Resnick in support of the Resnick Sustainability Institute. Three million dollars went to establish the Resnick Institute Innovation Fund to support clean energy research; the remaining $12 million went to a matching fund program for the Institute.[364] Also in May, Arizona State University received a $25 million grant from Julie Ann Wrigley, CEO of Wrigley Investments, to support research at ASU's Global Institute of Sustainability.[365] This, after she had already given $10 million in 2007. Then in August 2014, Columbia University scored $3.5 million from the Andrew Sabin Family Foundation

363 "The Crisis of Climate Disruption," American College and University Presidents' Climate Commitment. http://www. presidentsclimatecommitment.org/about/climate-disruption

364 "Caltech Receives $15 Million for Sustainability Research," *Philanthropy News Digest*, May 9, 2014. http://philanthropynewsdigest.org/news/caltech-receives-15-million-for-sustainability-research

365 "Julie Ann Wrigley, Facts and Figures," *Philanthropy News Digest*. http://philanthropy.com/factfile/gifts_detail?GiftDonorJoin_a_DonorID=PGDON1319

NAS

for Columbia's Center for Climate Change Law, which aims to fight land developers and other nature-despoilers in court.[366] In September 2014, the University of Arizona was given a $50 million bequest from the estate of Agnese Nelms Haury to support environmental and social justice research,[367] and the University of Dayton received a $12.5 million grant from the George and Amanda Hanley Foundation to establish a new sustainability institute.[368]

These high-dollar sustainability donations give some idea of how much money sustainability programs cost. Many Presidents' Climate Commitment signatories, however, are small regional colleges with lower profiles; likely they undertake these projects without significant grants.

Another telling data point is the proliferation of administrative sustainability jobs and their salaries. An average university sustainability director, according to a 2012 survey by the Association for the Advancement of Sustainability in Higher Education (AASHE), can expect to make approximately $82,791 per year. Energy managers scored the second-highest average salaries at $67,392, reaching as high in one case as $150,000. Sustainability managers averaged an annual wage of $62,059, assistant directors $60,345, education and outreach staff $48,658, recycling and waste staff $48,000, and sustainability coordinators $45,000.[369] An institution that hired staff for all seven positions could expect to spend upwards of $400,000 per year just on salaries alone.

366   "Columbia Law School Receives $3.5 Million for Climate Change Center," *Philanthropy News Digest*, August 20, 2014. http://philanthropynewsdigest.org/news/columbia-law-school-receives-3.5-million-for-climate-change-center

367   "University of Arizona Receives $50 Million Bequest," *Philanthropy News Digest*, September 24, 2014. http://www.philanthropynewsdigest.org/news/university-of-arizona-receives-50-million-bequest

368   "University of Dayton Gets $12.5 Million for Sustainability Institute," *Philanthropy News Digest*, September 24, 2014. http://www.philanthropynewsdigest.org/news/university-of-dayton-gets-12.5-million-for-sustainability-institute

369   Judy Walton, "Salaries and Status of Sustainability Staff in Higher Education—2012," Association for the Advancement of Sustainability in Higher Education, 2013, pg. 27. http://www.aashe.org/files/documents/programs/2012_staffsurvey-final.pdf.

Figure 9. Annual Salaries of Sustainability Staff[370]

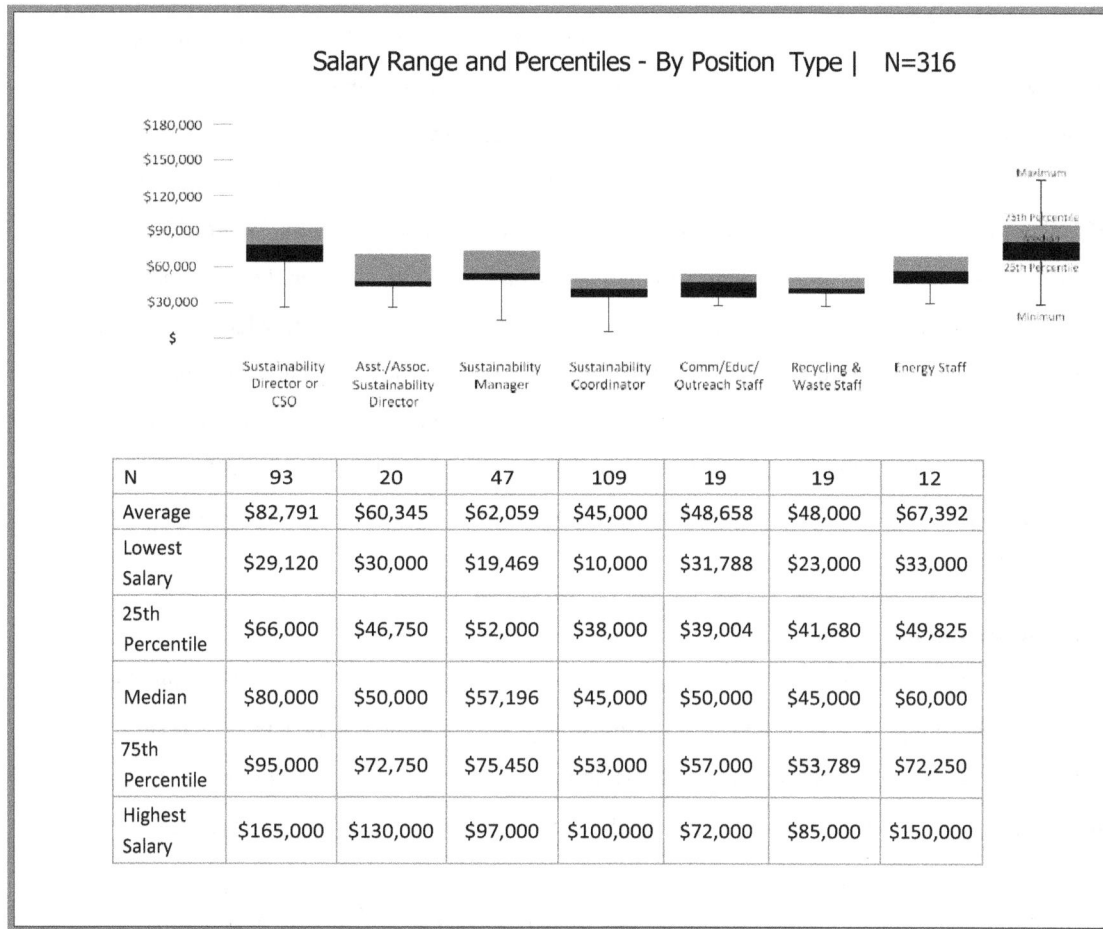

Salary Range and Percentiles - By Position Type | N=316

N	93	20	47	109	19	19	12
Average	$82,791	$60,345	$62,059	$45,000	$48,658	$48,000	$67,392
Lowest Salary	$29,120	$30,000	$19,469	$10,000	$31,788	$23,000	$33,000
25th Percentile	$66,000	$46,750	$52,000	$38,000	$39,004	$41,680	$49,825
Median	$80,000	$50,000	$57,196	$45,000	$50,000	$45,000	$60,000
75th Percentile	$95,000	$72,750	$75,450	$53,000	$57,000	$53,789	$72,250
Highest Salary	$165,000	$130,000	$97,000	$100,000	$72,000	$85,000	$150,000

In AASHE's survey, most institutions (82 percent) reported self-funding their sustainability staff salaries directly from the university's general operating fund, rather than paying the costs via gifts from external foundations and donors. AASHE also reported that "average salaries increased slightly across all regions and positions,"[371] and that 80 percent of all sustainability staff reported feeling that their positions were either "secure" or "very secure."[372] Even in times of rising student debt and shrinking college budgets, these sustainability officers were confident that their departments were high on the funding totem pole. Only 7 of the 450 surveyed (less than 1 percent) reported feeling "very insecure" about their position's funding. When AASHE asked the group about the biggest challenges facing their sustainability efforts, the most popular answer (50 percent) was lack of time to complete their goals; only 28 percent cited lack of funding.[373]

---

370   *Ibid*, pg. 27.

371   *Ibid*, pg. 27.

372   *Ibid*, pg. 32.

373   *Ibid*, 31.

NAS

What a complete sustainability department budget looks like is hard to say. Arizona State University in a 2013 annual report calculated that it had spent more than $5 million in a sustainability "revolving loan fund" alone.[374] These funds typically finance efficiency measures that should theoretically pay for themselves due to energy savings and utility rebates and other third-party incentives. The University of California's 2012 "Annual Report on Sustainable Practices" mentions receiving a $2 million grant to study whether renewable energy was effective at charging batteries for the university's electric cars, $66 million in energy efficiency grants, and $12 million in "incentive payments" from third parties who supported the renovation of 27 million square feet of building space to comply with energy efficiency programs sponsored by the local utility company.[375] Brown University acknowledges spending $14.6 million to reduce about 20,000 tons of carbon (saving, it says, about $3 million in energy reduction in the process).[376] But while Brown's report has plenty of statistics on the tons of pollution reduced, pounds of recyclables diverted, number of students involved, tallies of buildings retrofitted to higher efficiency standings, and the amount of electricity saved, there are no budgets, few dollar amounts, and no cost-benefit analyses.

Clashing Principles

The principle of marginal utility is a familiar concept: the more you acquire of a good, the less incremental value each unit brings to you. It works in the negative, too: reducing the first thousand tons of carbon is easier and less costly than reducing the last thousand. The purpose of a cost-benefit analysis is to determine at which point cost begins to outweigh the benefits—and to stop just before that point.

So why don't sustainability projects undergo more comprehensive public surveys of economic viability?

Sustainability operates on a different logical system—one that can fit within principles of marginal thinking, but that so reorders them as to be beyond much recognition. The main principle of decision-making within the environmental movement is something called the "precautionary principle." The term is a favorite of the EU and the UN and, given sustainability's birth in the UN's Brundtland Report, is an unsurprising kin to sustainability. The basic principle asserts that caution, rather than calculated risk-taking, should govern decision-making. If there is a chance that some action will cause harm, and no clear evidence or consensus among accepted experts as to whether it will or will not, then no action is to be preferred to

374 "SIRF: Sustainability Initiatives Revolving Fund, FY 2013 Annual Report," Arizona State University. http://www.asu.edu/pb/documents/SIRF-Annual-Report.pdf

375 "Annual Report on Sustainable Practices 2012," Budget and Capital Resources, University of California Office of the President, January 2013. http://sustainability.universityofcalifornia.edu/documents/Annual%20Report%20on%20 Sustainability%202012_+formatted_for_binder_v8.pdf

376 Christopher Powell, "Sustainability Progress Report," Office of Sustainable Energy and Environmental Initiatives, Facilities Management, Brown University, Fall 2012. http://brown.edu/Facilities/Facilities_Management/docs/Sustainability_Report_2012_Final.pdf

NAS

potentially risky action. Or if there is a chance that current activity could cause harm, even if there is no sure evidence of its doing so, one ought to scale back or even cease such activity and take measures to induce others to do so as well. In other words, until we know that global warming is not happening, or that greenhouse gas emissions do not exacerbate it, we should all aim to stop emitting greenhouse gases.

The European Court of Justice outlined the principle in the first legal case in which the precautionary principle was applied: "Where there is uncertainty as to the existence or extent of risks to human health, the institutions may take protective measures without having to wait until the reality and seriousness of those risks become apparent."[377] The trouble, as international trade lawyer Lawrence Kogan explains, is that "there is always some level of uncertainty, since certainty of the absence of risk is a logical, empirical, and scientific impossibility."[378]

Applied in its strongest, most literal sense, the precautionary principle would preclude most activities. Is there risk involved in driving a car? In eating sweets? In eating vegetables? Better to hold off and wait for better evidence. And because the precautionary principle favors weight of evidence over strength of evidence—preferring quantity over quality of relevant data—it discourages careful contemplation of facts and encourages quick conclusions drawn from surveys of the "relevant experts." According to Peter Saunders, professor emeritus of mathematics at King's College London, the precautionary principle is triggered once there is "at least prima facie scientific evidence of a hazard, rather than a risk."[379] Since it is impossible to prove the absence of some hazard, the outcome invariably is that the potential hazard is regulated or stopped. Hence, zealous zero-carbon policies that ignore standard economic analysis.

Without the precautionary principle, the sustainability movement would be hard-pressed to justify many of its initiatives. There is not enough evidence to marshal in favor of radical carbon-cleanses. Only the principle of precaution, rather than careful cost-benefit risk analysis, could value carbon so highly as to make it worth exorbitant costs to remove. Measured by this principle, in an environmentalist Pascal's wager, eliminating carbon emissions makes sense. There is a slight chance—however slim—that minuscule amounts of carbon contribute to runaway global warming, which itself has a chance of harming us. Follow the logic back and the decision is clear: quit emitting, quit trashing, quit polluting. At that point, no price is too high.

---

377   Cited in Lawrence Kogan and Lucas Bergkamp, "Trade, the Precautionary Principle, and Post-Modern Regulatory Process," *European Journal of Risk Regulation*, April 2014, pg. 498.

378   Kogan, pg. 498.

379   Cited in Kogan, pg. 499.

NAS

Consider Middlebury College's Carbon Neutrality Plan. The plan does not specifically name the precautionary principle as its guiding standard, but it uses the language of potential unknown hazards as justification for its carbon neutral goal. The preface praises the Pew Center on Global Climate Change for serving "as an influential forum for corporate and non-profit leaders to objectively explore the environmental and economic risks and uncertainties associated with climate change"[380] and devotes an entire section, I.2.1, to "Potential climate impacts on Vermont."[381] Here the college acknowledges that

> *Admittedly, the uncertainties associated with regional predictions of the consequences of climate change are higher than the uncertainty in global predictions.*[382]

But "despite these uncertainties," Middlebury is preparing for a host of possible detrimental outcomes predicted by the EPA. It lists a series of risks that it hopes to mitigate with its carbon neutrality plan: a 4-degree Fahrenheit temperature rise by 2100; an increase in rainfall; heat-related illnesses among the elderly; concentrations of ozone leading to eye irritations, asthma and other ailments; upticks in tick-borne diseases; shortened ski seasons; the displacement of maple trees by oak and hickory; a change in fall leaf colors; and the disruption of the maple syrup industry.[383]

Planning for possible dangers is wise of course, but so is considering possible benefits that may come from warming temperatures: longer growing seasons, greater flora fertility thanks to increased carbon, more warm-season tourism, less road salt damage, or lower heating bills. Cost-benefit analyses consider both sides of these equations; the precautionary principle considers only the potential harms.

The Story of Middlebury

We offer a case study of one college that has positioned itself as a sustainability icon. Middlebury College is known for its commitment to achieving carbon neutrality. Bill McKibben, probably the best-recognized environmental thinker and activist in the country, is a resident scholar there.

Middlebury College's zero-carbon goal has a long history behind it. In 1999, well before most of its peer institutions began devoting themselves to environmental causes, and seven years before Al Gore's documentary *An Inconvenient Truth* conveniently stirred up public fear of climate change, the college had identified global warming as a significant threat it pledged to fight. Two years later, in a 2001 "Environmental Peak Report," a task force of professors set forth 140 action items meant to make the pledge a reality. One of these items was a recommendation to "establish a carbon neutral campus with

---

380   *Carbon Neutrality at Middlebury College*, pg 9.

381   *Ibid.*

382   *Ibid.*

383   *Ibid*, pg. 10.

NAS

zero emissions"—though no dates were set and no vows sworn just yet.[384]

Students took up the cause. One particularly dedicated group signed up for a January 2003 winter term course on the "Scientific and Institutional Challenges of Becoming Carbon Neutral." Their coursework involved researching case studies and writing a 200-page brief on strategies Middlebury might adopt to reduce its carbon emissions to a net of zero.[385] Their efforts convinced the board of trustees to vote in 2004 to reduce emissions by 8 percent below 1990 levels by the year 2012. Two years later, after more student research and outreach, including another winter term course that hosted a "MiddShift" climate change conference in January 2006, the board upped those goals to full climate neutrality by 2016. The official proclamation tied carbon neutrality to the college's deepest values, and linked the project to sustainability:

> *We believe the College should take a leadership stance on carbon neutrality and should build and expand upon the strategies it has in place to attain carbon neutrality and take further actions to develop and implement sound strategies that ultimately advance sustainability for this institution and our planet.*[386]

One year later President Ronald Liebowitz made the pledge a matter of public accountability when he joined 151 college presidents as charter signatories of the newly launched American College and University Presidents' Climate Commitment.[387]

The role of the Biomass Gasification Plant in this endeavor was to cut down Middlebury's reliance on No. 6 fuel oil, the main source of energy the college had previously used to heat and to supply electricity around campus. The students in the winter term course on carbon neutrality calculated that the oil burned in campus heating equipment constituted the single greatest source of greenhouse gas emissions on campus. By installing a 30,000-pound biomass boiler, Middlebury could cut down 1 million gallons of oil— 50 percent of its annual usage—and 40 percent of its annual carbon emissions.[388] When the plant

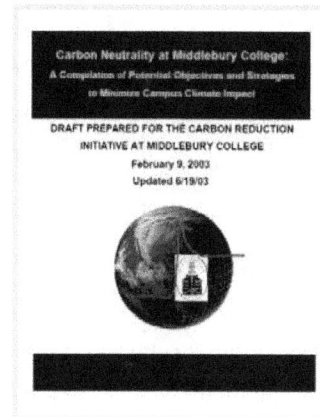

---

384   Nan Jenks-Jay and Chris McGrory Klyza, "Dear Alumni and Friends," *Environmental News*, Third Issue, Spring 2002, pg. 3. http://www.middlebury.edu/media/view/101461/original/EnvNews02.pdf.

385   *Carbon Neutrality at Middlebury College: A Compilation of Potential Objectives and Strategies to Minimize Campus Climate Impact*, February 9, 2003. https://www.middlebury.edu/media/view/262585/original/es010_report.pdf

386   "Resolution on Achieving Carbon Neutrality by 2016," Middlebury College, May 5, 2007. http://www.middlebury.edu/ sustainability/policy-planning/policies/neutrality/2007

387   "Charter Signatories," American College and University Presidents' Climate Commitment, 2007. http://www2. presidentsclimatecommitment.org/html/chartersignatories.php

388   *Carbon Neutrality at Middlebury College*.

opened in frigid February 2009, Middlebury could declare itself well on its way to full carbon neutrality.

In fact, however, the plant provoked disagreement from those skeptical that biomass could truly be carbon neutral. In 2006, immediately after Middlebury announced its interest in pursuing biomass, the *Chronicle of Higher Education* published "Truth in Advertising: Middlebury College's Biomass Plant," which noted that saw mills (themselves consumers of energy) would produce the woodchips, that Middlebury's entry into the chip market would increase demand for wood, and that the net result might be a greater number of trees cut down and therefore less carbon sequestered.[389]

Middlebury faculty and students also began to dispute that biomass could truly be carbon-neutral. During the fall 2009 semester, Christopher Klyza, the Stafford Professor of Public Policy, Political Science, and Environmental Studies, taught a course on Middlebury's biomass plant in which students investigated whether biomass boilers actually released carbon emissions. "The students were interested in this question, because it didn't make sense that there is smoke coming out of the biomass plant," Klyza told the student newspaper, *The Middlebury Campus*. "It's not obviously carbon neutral. So there must be more to it."[390] Even Jon Isham, Director of the Center for Social Entrepreneurship and Professor of Economics, who in 2008 taught an environmental economics course in which students researched and first recommended the biomass plant to the college administration, had to admit in 2013, "I think we've rethought biomass and how carbon neutral it is. There were some critiques from faculty colleagues that proved to be true about overselling biomass as a carbon neutral process."[391] Middlebury, for its part, insists that it is responsible only for emissions from its own geographic property or from its private operations, not including any emissions from chopping or transporting the wood, carting off the ashes, or other third-party involvement.

(Bio)Mass Budget Destruction

Middlebury College, like most, does not publicize its sustainability budget. Indeed, it doesn't seem to have one per se. The costs of sustainability are diffused through many departmental budgets and are not broken out even within those budgets. But a great deal of document sifting and researching yields several startling conclusions: Middlebury appears to underestimate costs and overestimate benefits of sustainability endeavors, and the college appears driven more by ideology than by principles of economics in undertaking new sustainability programs.

---

389   Richard Montastersky, "Truth in Advertising: Middlebury College's Biomass Plant," *Chronicle of Higher Education*, October 20, 2006. http://chronicle.com/article/Truth-in-Advertising-/27972

390   Claire Abbadi, "Carbon Neutral, or Carbon-Lite?" *The Middlebury Campus*, October 16, 2013. http://middleburycampus.com/article/carbon-neutral-or-carbon-lite/

391   Abbadi, "Carbon Neutral, or Carbon-Lite?"

NAS

We recognize that individual institutions—especially private ones—are free to allocate their budgets as they see fit. Middlebury College has a right to spend abundantly on sustainability, and it is evidently willing to do so. Our aim in producing a case study of Middlebury College's sustainability costs is solely to offer an estimate—the first we have seen to date—of what campus sustainability initiatives cost. To this end we include costs that might in themselves seem trivial but as components of overall expense are worth registering. We also recognize that the expenses that Middlebury College or any other college voluntarily assumes have to be weighed against the social good that these institutions imagine they are contributing to. In this analysis, we aggregate the costs and take account of any savings Middlebury reports, but we make no attempt to quantify the more diffuse social benefits that these expenditures are intended to bring about.

The new biomass plant at Middlebury is a good example of expensive sustainability-minded projects. According to a notice on the Planning, Design, and Construction division of Middlebury's Facilities Office, the new biomass-based heating system cost $11.9 million to construct,[392] though L.N. Consulting, Inc., the company that designed and oversaw the construction of the plant, pegged the cost at $12.5 million. [393] The college explains in a report to the Presidents' Climate Commitment that it financed that expense on the basis of student fees, outside grants and individual contributions, and money it borrowed through a bond sale.[394]

The plant is among Middlebury's pricier campus expenditures. The Planning, Design, and Construction division records that the same year that the plant opened, in 2009, Middlebury spent $5 million renovating the student center[395] and $1.7 million on a ski lift at the snow bowl.[396] Its most expensive campus construction that year was a $10.6 million upgrade on the dining hall.[397] The biomass plant cost the equivalent of what 263 students (about 11 percent of the student body) will spend on tuition for the 2014-2015 school year ($45,637 each),[398] or, for comparison, about 4 percent of the fiscal year 2013 operating budget—not too far behind the library services at 6 percent.[399]

---

392   "Biomass Gasification Facility," Middlebury College. http://www.middlebury.edu/media/view/185271/original/Biomass_Project.pdf

393   "Middlebury College Biomass Project," L.N. Consulting, Inc., 2009. http://www.lnconsulting.com/portfolio/plant-systems/middlebury-college-biomass-project

394   "Progress Report for Middlebury College," ACUPCC Reporting System.   http://rs.acupcc.org/progress/390/

395   "McCullough Student Center," Middlebury College. http://www.middlebury.edu/media/view/185321/original/McCullough_Project.pdf

396   "Worth Mountain Chair Lift at the Snow Bowl," Middlebury College. http://www.middlebury.edu/media/view/189061/original/Worth_Mountain_Project.pdf

397   "Proctor Dining Hall," Middlebury College. http://www.middlebury.edu/media/view/185351/original/Proctor_Project.pdf

398   Frequently Asked Questions, Middlebury Admissions.   http://www.middlebury.edu/admissions/start/faq/node/172721

399   Frequently Asked Questions, Middlebury Financial Facts. http://www.middlebury.edu/media/view/464848/original/financial_faqs_17_infosheet_12.13.pdf

NAS

The boiler system, itself a $2.5 million-dollar[400] piece of equipment from Chiptec, Inc. of Bristol and Williston, Vermont,[401] is housed in an 8,000 square-foot addition to an existing service building, about 5,000 square feet of which also had to be renovated. Its operation requires a staff of six, as well as an electronic video-based control system.

To feed the boiler, the college buys wood from local farms and forests. In an experiment co-run with the State University of New York College of Environmental Science and Forestry, Middlebury Professor Timothy Volk had attempted to grow wood for the engine's consumption, planting nine acres of willow trees on the west side of Middlebury's campus to be harvested on a rotating three-year basis. Willow, a fast-growing, perennial tree that grows to sufficient height within three years, seemed ideally suited to the college's needs for quick wood that could be planted and harvested easily by machine. But seven years' testing (two harvests) determined the wood was too wet, too cold, and insufficient in quantity to supply the biomass needs of the plant, and the college would need to continue purchasing wood from outside sources.

The idea for biomass surfaced in the 2003 student report as an alternative to biodiesel fuel to replace the No. 6 fuel oil. Biodiesel, at $1.30 per gallon, was too expensive to replace the carbon-heavy fuel oil, whereas biomass—wood chips gathered from sawmills, forest floors, or chopped trees—was by comparison cheap. A consultant team with Vermont Family Forests determined that the surrounding Addison and Rutland counties had an annual supply of 269,250 green tons of suitable low-quality wood, and about 109,592 annual green tons in current demand for firewood, wood pulp, and wood chips, leaving 159,658 green tons of wood per year available for potential use, dependent on whether the landowners were interested in cutting and selling.[402]

Though the biomass chips were less expensive than biodiesel, the construction project wasn't cheap to undertake. In their initial estimate, the students guessed that the biomass gasification plant would cost $2.52 million in start-up costs: $2 million for the biomass boiler itself, $500,000 for delivery and storage equipment, and $23,000 for a storage building to house wood chips.[403] It would last for 50 years. But on the other hand, by reducing the college's need for fuel oil, it could save Middlebury $309,300 in annual

400   Michele Madia and Tadu Yimam, "How Many Wood Chips Does a Wood-Chip Burner Burn?" *Business Officer Magazine*, National Association of College and University Business officers, June 2010. http://www.nacubo.org/Business_Officer_Magazine/Magazine_Archives/June_2010/Energy_Stewards.html

401   "Chiptec Wood Energy Systems Helps Middlebury College Become Carbon Neutral," Chiptec Wood Energy Systems. http://www.chiptec.com/linked/middlebuiry%20college%20case%20study%20final.pdf

402   "Biomass Fuel Assessment for Middlebury College," Vermont Family Forests, January 31, 2004, pg. 44. http://sites.middlebury.edu/biomass/files/2009/02/mcbiomassreport.pdf

403   *Carbon Neutrality at Middlebury College*, pg. 37.

NAS

fuel costs.[404] In 7 years, they thought, it would pay for itself.

There were ancillary expenses, though: an estimated $250,000 in annual operating costs (which they thought would match the current operating costs for the fuel oil boiler), $5,500 per year to landfill the resulting ash (unless Middlebury expanded its compost site, or succeeded in selling the ash as fertilizer), plus about $25 per ton of wood chips (at 20,000 tons, about $500,000 per year).[405]

The actual cost of construction, when it began four years later, turned out to be almost five times as expensive, and the cost of woodchips rose to $37 per ton. (Meanwhile, the cost of No. 6 fuel oil also rose, from $0.80 to $1.50 per gallon.)[406] The expected life of the plant was not actually 50 years, but much lower, somewhere in the range of 25 to 30. And, while the college expects to save $840,000 in annual fuel costs, the anticipated payback period is now 12 years, introducing more   risk.[407]

"Pay for itself" is a misleading phrase, though. In its angst to save carbon and label the project budget-viable, Middlebury appears to have omitted various costs. Does the savings of $840,000 per year in the cost of biomass relative to the cost of No. 6 fuel oil factor in the costs of operating the plant? The students in 2003 estimated $250,000 to staff and maintain the plant (about $323,709 in 2014 dollars, according to the Bureau of Labor Statistics), which it thought would match previous years' expenditures on operating the No. 6 fuel oil boiler. But an analysis of Middlebury documents show that merely paying the plant's operating staff easily tops that $250,000 figure.

A 2014 document from Middlebury's Human Resources Department listing all staff positions shows that Middlebury has hired five "Heating Plant Operators" to run the biomass plant, and one "Manager of the Central Heating Plant." Kelly Boe, the heating plant manager, told us in an interview that there are actually six, not five plant operators. The exact salaries of these staff members are private, but the document does list the "salary band levels" of all staff, which refer to specific salary ranges.[408] Another document from July 2014 lists the average pay ranges from the lower, middle, and upper thirds of the staff members within these bands, with slight differences within bands based on seniority, performance, and other individual factors.[409] Cross-checking the documents and averaging the salary data yield a reasonably

---

404   *Carbon Neutrality at Middlebury College*, pg.   38

405   *Carbon Neutrality at Middlebury College*, pp.   37-38.

406   *Carbon Neutrality at Middlebury College*, and FAQs, Biomass at Middlebury. http://sites.middlebury.edu/biomass/about/faqs/

407   FAQs, Biomass at Middlebury.

408   "Staff Positions," Middlebury College, Department of Human Resources, January 20, 2014. http://www.middlebury.edu/media/view/355943/original/staff_positions_by_bandlevel.pdf

409   "Middlebury College Staff Pay Ranges," Middlebury College, Department of Human Resources, July 1, 2014. http://www.

accurate portrayal of what Middlebury actually pays these six staff for the biomass gasification plant: approximately $87,000 annually for the manager, and approximately $42,000 per year for each of the six operators, for a total of just over $338,000.

Figure 10. Estimated Annual Salaries of Staff for Middlebury's Biomass Gasification Plant

Position	Salary Band	Lowest Salary in Band	Highest Salary in Band	Average Salary in Band
Manager of the Central Heating Plant	Management 3	$63,170	$110,543	$86,857
Heating Plant Operator	Specialist 1	$31,595	$52,128	$41,862
Heating Plant Operator	Specialist 1	$31,595	$52,128	$41,862
Heating Plant Operator	Specialist 1	$31,595	$52,128	$41,862
Heating Plant Operator	Specialist 1	$31,595	$52,128	$41,862
Heating Plant Operator	Specialist 1	$31,595	$52,128	$41,862
Heating Plant Operator	Specialist 1	$31,595	$52,128	$41,862
Total		$252,740	$423,311	$338,029

Another 2014 document summarizing the health, dental, and vision insurance, and retirement contributions that Middlebury offers also gives some indication of the benefits these seven staff likely receive.[410] Middlebury pays a variable percentage of health insurance based upon salary level (the higher the salary, the lower the percentage), but a fixed portion of dental and vision insurance. Assuming that each staff member has opted for a two-person insurance coverage plan (rather than single, family, or none at all), and basing the percentages on the estimated salary ranges, we estimate that Middlebury spends approximately $83,000 on these seven staff members' health insurance, $5,700 on dental insurance, and $470 on vision insurance, for a total of about $89,000 in insurance benefits.

---

middlebury.edu/media/view/476423/original/2014_middlebury_staff_pay_ranges.pdf

410 "Middlebury College Benefits Summary 2014," Middlebury College, 2014. http://www.middlebury.edu/media/view/468704/original/midd_-_benefits_summary_2014.pdf

NAS

Figure 11. Estimated Annual Insurance Benefits of Staff for
Middlebury's Biomass Gasification Plant

Position	Health Insurance	Dental Insurance	Vision Insurance	Total Cost of Insurance
Manager of the Central Heating Plant	$11,142	$814	$67	$12,023
Heating Plant Operator	$12,007	$814	$67	$12,888
Heating Plant Operator	$12,007	$814	$67	$12,888
Heating Plant Operator	$12,007	$814	$67	$12,888
Heating Plant Operator	$12,007	$814	$67	$12,888
Heating Plant Operator	$12,007	$814	$67	$12,888
Heating Plant Operator	$12,007	$814	$67	$12,888
Totals	$83,184	$5,698	$470	$89,351

Middlebury's retirement plan requires that all employees over the age of 21 set up and contribute towards an account, so it is reasonable to assume that all seven biomass plant staff have such an account. The college has a progressive retirement account that rewards longer tenures with greater contributions and that also gives larger contributions towards older staff members' accounts. Middlebury contributes 3 percent of each staff member's salary for those who have worked for the College 0-2 years, 9 percent for those who have worked there 2 or more years and are between the ages of 21 and 44, and 15 percent for those who have worked for Middlebury at least 2 years and who are 45 years or older.[411] Taking the middle estimate, 9 percent, yields retirement contributions of about $7,800 for the plant manager and $3,800 for each of the six operators, for a total of approximately $30,000 in retirement contributions.

Figure 12. Estimated Annual Retirement Benefits of Staff for Middlebury's
Biomass Gasification Plant

Position	Average Salary in Band	Retirement
Manager of the Central Heating Plant	$86,857	$7,817
Heating Plant Operator	$41,862	$3,768
Heating Plant Operator	$41,862	$3,768
Heating Plant Operator	$41,862	$3,768
Heating Plant Operator	$41,862	$3,768
Heating Plant Operator	$41,862	$3,768
Heating Plant Operator	$41,862	$3,768
Totals	$338,029	$30,425

---

411 "Middlebury College Benefits Summary 2014."

NAS

Counting salaries and benefits, we estimate that Middlebury spends approximately $457,000 compensating its biomass plant workers—about $133,000 above the $323,709 inflation-adjusted figure that the students had estimated as the annual total cost of operating the old fuel oil boiler and had projected as the new annual cost of operating the biomass boiler. (The students' estimate came without a breakdown of component costs, calculations, or sources of information.) This increase in costs is in keeping with other case studies. When the University of Minnesota-Morris in 2008 installed a biomass gasification plant a little less than half the capacity of Middlebury's (it requires 9,000 tons[412] of biomass per year, relative to Middlebury's 20,000), the university estimated that it spent an additional $132,600 in labor costs, beyond what it had spent previously to man its gas-based boiler.[413]

Figure 13. Estimated Annual Compensation of Staff for Middlebury's Biomass Gasification Plant

Position	Salary Band	Average Salary in Band	Estimated Benefits	Estimated Total Compensation
Manager of the Central Heating Plant	Management 3	$86,857	$19,840	$106,697
Heating Plant Operator	Specialist 1	$41,862	$16,656	$58,518
Heating Plant Operator	Specialist 1	$41,862	$16,656	$58,518
Heating Plant Operator	Specialist 1	$41,862	$16,656	$58,518
Heating Plant Operator	Specialist 1	$41,862	$16,656	$58,518
Heating Plant Operator	Specialist 1	$41,862	$16,656	$58,518
Heating Plant Operator	Specialist 1	$41,862	$16,656	$58,518
Total		$338,029	$119,776	$457,805

The $457,000 gross figure does not include any maintenance, repairs, or other operational costs ofactually running the boiler. These are non-trivial costs for which Middlebury supplies no information. Additional annual operating expenses include ash removal. Boe, the plant manager, told us that Middlebury stores its ash in an underground bunker, and four times each year pays a private company to remove the ash and turn it into fertilizer for local farms. Boe did not know the cost of the ash removal, while Middlebury's website merely notes that "Ash is collected and used by a local fertilizer company in their products."[414]

Maintenance constitutes another major cost. Most biomass systems require constant observation, and

412   Judy Riley, "Public Dedication of Biomass Gasification Facility," Morris Campus News and Events, October 3, 2008. http://www.morris.umn.edu/newsevents/view.php?itemID=6547

413   Joel Tallaksen, Arne Kildegaard, "Chapter 4: Financial and Economic Analysis," Final Report to the USDA Rural Development Grant 68-3A75-5-232. University of Minnesota-Morris, pg 18, Table II. http://renewables.morris.umn.edu/biomass/documents/USDA_Report/SII_Finance.pdf

414   "What Happens to the Ash?" FAQs, Biomass at Middlebury, Middlebury College. http://sites.middlebury.edu/biomass/about/faqs/

NAS

periodically equipment such as fans, motors, conveyers, pressure parts, water wall tubes, super heaters, and other parts need replacement.[415] In 2011, the baghouse filter system, which purifies the air emitted through the smokestack of any particulates, caught fire spontaneously and needed to be completely replaced.[416] Biomass plants at nearby Green Mountain College and the town of Poultney had also faced malfunctioning filtering systems.

And every couple of months, the system must be shut down, cooled off, and thoroughly cleaned of ash, charcoal, mineral residue, and any debris that gets carried along with the wood.[417] The process takes several days. Ideally that cleaning should take place every few months, says plant manager Kelly Boe, but during October 2013, the biomass team pushed the plant to run for a record 16 weeks without cleaning.[418] Eventually, the college would like to run the plant for a year straight, Boe told us, but for now they're working towards the intermediate goal of only four planned shut-downs per year. The reason for delaying the scheduled cleaning is to avoid additional carbon emissions. When the biomass plant shuts down, the campus runs on fuel oil instead. "It is painful for us to use oil," Boe commented to Middlebury's student newspaper, *The Middlebury Campus*. "No one wants to be the guy that breaks the streak" of going oil-free.[419] "The really significant part of [running for a consecutive 16 weeks] is that it means we burn that much more biomass and that much less fuel oil," commented Jack Byrne, the Director of Sustainability Integration.[420]

Middlebury does not reveal its annual maintenance costs, but a representative from Chiptec Inc., the manufacturer of Middlebury's biomass plant, told us in an interview that it often recommends that customers plan to spend about 2 percent of the cost of their initial infrastructure investment on annual maintenance. In Middlebury's case, 2 percent of the upfront $2.5 million to purchase the boiler system comes to $50,000 per year.

As a corollary, consider the University of Minnesota-Morris case study once again. UMM estimated that it needed to purchase new equipment (a telehandler, semi-truck with walking floor trailer, conveyer, and

415   Anna Austin, "Planned Outage Protocol," *Biomass Magazine*, January 7, 2013. http://biomassmagazine.com/articles/8491/planned-outage-protocol

416   Susie, Steimle, "Middlebury Biomass Fire Still a Mystery," *WCAX News*, May 20, 2011. http://www.wcax.com/story/14687547/middlebury-biomass-fire-still-a-mystery

417   Joe Flaherty, "Round the Clock, Selleck Runs Biomass Plant," *The Middlebury Campus*, October 16, 2013. http://middleburycampus.com/article/round-the-clock-selleck-runs-biomass-plant/

418   Joe Flaherty, "Carbon Cleanup," *The Middlebury Campus*, October 10, 2013. http://middleburycampus.com/article/carbon-cleanup-an-inside-look-at-what-happens-when-the-biomass-plant-shuts-down/

419   Mitch Perry, "Biomass Plant Cuts Oil Use By 600,000 Gallons," *The Middlebury Campus*, October 9, 2013. http://middleburycampus.com/article/biomass-plant-cuts-oil-use-by-600000-gallons/

420   *Ibid.*

NAS

tractor with a bucket loader) approximately every ten years at a total cost of $180,000, or about $18,000 per year. The additional cost of operating this equipment was estimated at $8,000. Other maintenance costs included cleaning, repairing, and replacing various parts (sensor, filters, the boiler, the refractory lining of the gasifier) at about $27,000 per year. And the biomass plant required additional supplies: $15,000 in sodium hydroxide (NaOH) and $5,000 in additional water. Per year, these maintenance costs totaled $73,000, in addition to the extra labor involved.[421]

Figure 14. Estimated Annual Maintenance Costs for the Biomass Plant at the University of Minnesota-Morris[422]

Table III- Operations and Maintenance For Biomass Plant

Predicted Annualized Auxiliary Equipment Costs
(Equipment for biomass handling logistics)

Telehandler	50,000
Semi-Truck with walking floor trailer	70,000
Conveyer	10,000
Tractor With Bucket Loader	50,000
(purchase price for used equipment- 10 yr life)	$ 180,000.00

Equipment Capital	$ 18,000
Equipment Operational	$ 8,000
Sub-Total	$ 26,000

Predicted Additional Maintenance
(Not including Plant staff labor)

Estimated Total Expenses	$ 15,000
Parts Boiler Cleaning	
Sensors Filters	
Refractory Maintenance	$ 12,000
Sub-Total	$ 27,000

Additional supplies

NaOH	$ 15,000
Water	$ 5,000
Electricity	
Sub-Total	$ 20,000

Total Yearly	$ 73,000

Another expense Middlebury pays is for the wood to feed the plant. Boe estimates that the plant requires 25,000 tons of wood chips each year, but declined to name a wood chip budget. Elsewhere, the college estimates a price of $37 per ton, which by our calculation makes for a total of $925,000 per year.[423]

421   Tallaksen and Kildegaard, "Chapter 4: Financial and Economic Analysis." Table III, pg. 19.

422   *Ibid.*

423   "How Much Did the Biomass Project Cost and What Is the Payback?" FAQs, Biomass at Middlebury, Middlebury College.

NAS

Middlebury estimates slightly lower and puts a positive spin on this expense: "The project will also pump $800,000 annually into the local economy through the purchase of woodchips."[424] (Relative to the cost of No. 6 fuel oil, Middlebury believes this saves $840,000 in fuelcosts.)[425]

There are other questions left unanswered. Did the college price the research and time required to find the proper wood and to determine that willow didn't work? Do the projected savings assume peak operating efficiency, which presumably will decline over time, and already has declined due to less-than-ideal wood supplies?

What about the staff time required to research and vet the proposal, and the hours of work to calculate how many tons of carbon the plant would save, and whether those savings would suffice to advance Middlebury's crusade towards carbon neutrality? At various points, the preparation costs for the biomass system were kept artificially low by turning biomass research—that might have required a consultant or staff time at the Office of Sustainability—into student homework assignments. Not only was the 200-page carbon neutral strategy that recommended biomass boilers written by students in a 2003 winter term, but Middlebury's first emissions inventory that laid the groundwork for climate neutrality by tracking all of the college's sources and volumes of greenhouse gas emissions was conducted as part of a student internship and senior thesis in 2001.

And the 2008 Climate Action Plan in which Middlebury laid out its official strategy to achieve carbon neutrality, submitted for review to the Presidents' Climate Commitment, drew on the senior theses and independent studies of six different undergraduate students. Additionally, for answers to key questions— such as *What is the availability of biomass fuel to replace 1 million gallons of No.6 fuel oil? What is the local, regional, and global economic impact of Middlebury College procuring 1.2 million gallons per year of a B100 biodiesel fuel source?*—the Climate Action Plan cited student work in Professor Jon Isham's Spring 2008 Environmental Economics course.[426]

In other cases, the numbers appear evasive. The estimated cost savings of $840,000 per year, divided into the upfront costs of $11.9 million, equals more than 14 years to recover the initial investment—not 12, as Middlebury's documents claim. This indicates that third parties are covering substantial costs that do not figure into Middlebury's internal calculations. Indeed, Middlebury's 2012 Progress Report

http://sites.middlebury.edu/biomass/about/faqs/

424   *Ibid.*

425   *Ibid.*

426   MiddShift Implementation Working Group, "Climate Action Implementation Plan," Middlebury College, August 28, 2008, pp 27-28. http://www.middlebury.edu/media/view/243071/original/Middlebury_CAP.pdf

NAS

to the American College and University Presidents' Climate Commitment discloses that it had received $1,476,893 in outside grants and gifts to be used for sustainability. The report does not detail what Middlebury spent these gifts on, but if the sum went towards the biomass plant, the remaining cost to Middlebury ($10,423,107) is equal to just over 12 years' savings at $840,000 per year.

Private donations and grants are normal parts of university initiatives, and undoubtedly the donors' intentions are good. But whereas they make a biomass gasification plant more affordable to Middlebury, they do not make the plant more economically viable overall. That is, Middlebury might (and one must emphasize the speculative nature of that "might") privately gain from its subsidized construction of a biomass plant, but the broader society and economy as a whole might experience a loss in the form of an inefficient use of funds.

One has to wonder as well whether the biomass boiler will remain in use for the full duration of its expected 25-year lifetime. If a new carbon-neutral system less reliant on trees is invented, will the college switch in an attempt to update its environmental profile?

The Wages of Sustainability

Middlebury's sustainability efforts go far beyond its biomass plant. An entire team of staff works on carbon-reduction projects and other environmental initiatives. Middlebury has a Dean of Environmental Affairs, a Director of the School of Environment, a Director of the Sustainability Integration Office, a Sustainability Communications/Outreach Coordinator, a School of Environment Language School Personnel and Budget Coordinator, another School of Environment staff member, an Assistant Director of the Franklin Environmental Center, a Food and Farm Educator to run the campus organic farm, an Environmental Health and Safety Coordinator, a Coordinator for Community Environmental Studies, and one Ecologist on staff—in addition to the biomass plant staff.

Using the same documents used to estimate the salaries of the biomass plant staff, we estimated the salaries of these positions. Four positions—the Dean of Environmental Affairs, the Director of the School of Environment, the School of Environment staff, and an ecologist— did not include a salary band level. We estimated these salaries using other sources documented below.

Figure 15. Estimated Annual Salaries of Sustainability Staff at Middlebury College

Position	Salary Band	Lowest Salary in Band	Highest Salary in Band	Average Salary in Band
Dean of Environmental Affairs	not graded			$197,475*
Director of Sustainability Integration Office	Administrator 1	$75,341	$139,357	$107,349
Sustainability Communications/Outreach Coordinator	Specialist 1	$31,595	$52,128	$41,862
Assistant Director of the Franklin Environmental Center	Specialist 3	$46,305	$76,409	$61,357
Food and Farm Educator	Specialist 3	$46,305	$76,409	$61,357
Environmental Health and Safety Coordinator	Specialist 4	$57,851	$95,471	$76,661
Director of the School of Environment	not graded			$118,670**
School of Environment Staff	not graded	$30,668†	$49,065†	$39,867†
School of Environment Language School Per Budget Coordinator	Operations 4	$39,353	$62,969	$51,161
Coordinator Community Environmental Studies	Specialist 3	$46,305	$76,409	$61,357
Ecologist	not graded	$57,851‡	$95,470‡	$76,661‡
Total				$893,777

* The dean's salary was estimated based on the average salary of a dean of Forestry and Environmental Studies, according to the 2012-2013 Administrators in Higher Education Salary Survey. [427]

** The salary for the director of the school of the environment was estimated based on the average salary of an assistant dean of Forestry and Environmental Studies, the closest match in the College and University Professional Association for Human Resources, 2012-2013 study of higher education administrators' salaries.

† These numbers for the School of Environment Staff member were calculated on the assumption that the position, though not assigned to a salary band, might resemble a level-3 operations position, as other jobs in this band include various secretarial and programmatic work.

‡ These numbers for the Ecologist were calculated assuming that the position is paid at the same rate as a specialist level 4 (the highest level of specialists). Because faculty are not included on this staff list, we assume that the Ecologist, though likely holding a Ph.D., is not paid at the rate of Middlebury professors.

Adding in insurance and retirement benefits brings the total compensation to over $1.1 million.

---

427   *Administrators in Higher Education Salary Survey*, College and University Professional Association for Human Resources, 2012-2013, pg 20. http://www.cupahr.org/surveys/files/salary2013/AHE13-Executive-Summary.pdf

NAS

Figure 16. Estimated Total Annual Compensation of Sustainability Staff at Middlebury College

Position	Average Salary in Band	Health Insurance	Dental Insurance	Vision Insurance	Retirement	Total Compensation
Dean of Environmental Affairs	$197,475	$10,085	$814	$67	$17,773	$226,214
Director of Sustainability Integration Office	$107,349	$10,680	$814	$67	$9,661	$128,571
Sustainability Communications/ Outreach Coordinator	$41,862	$12,007	$814	$67	$3,768	$58,518
Assistant Director of the Franklin Environmental Center	$61,357	$11,574	$814	$67	$5,522	$79,334
Food and Farm Educator	$61,357	$11,574	$814	$67	$5,522	$79,334
Environmental Health and Safety Coordinator	$76,661	$11,574	$814	$67	$6,899	$96,015
Director of the School of Environment	$118,670	$10,680	$814	$67	$10,680	$140,911
School of Environment Staff	$39,867	$12,440	$814	$67	$3,588	$56,776
School of Environment Language School Per Budget Coordinator	$51,161	$12,007	$814	$67	$4,604	$68,653
Coordinator for Community-Based Environmental	$61,357	$11,574	$814	$67	$5,522	$79,334
Ecologist	$76,661	$11,574	$814	$67	$6,133	$95,249
Total	$893,777	$125,769	$8,954	$737	$79,672	$1,108,909

There are also three Waste & Recycling Handlers (each paid approximately $31,625, plus benefits), and one Waste & Recycling Hauler (approximately $39,867 plus benefits) whose jobs presumably would be needed with or without sustainability initiatives. We did not include these jobs in our tally.

NAS

This list of administrative and operational positions does not include the staff for the biomass gasification plant.

Nor does it include any of the environmental and sustainability faculty members: 1 instructor, 2 lecturers, 2 assistant professors, 1 associate professor, 2 full professors, plus 7 other professors of psychology, economics, English, and various other disciplines who spend part of their teaching assignments on environmental courses, and 10 senior fellows with Middlebury's summer program, the School of the Environment. An internal study comparing Middlebury's average faculty salaries with those of other institutions in 2012-2013 shows that an average Middlebury assistant professor ranked seventh among 12 elite liberal arts college, receiving about $101,409 each year (including benefits). An average Middlebury associate professor receives $119,429 (ranking eighth among the 12 colleges). And a full professor at Middlebury can expect to receive approximately $172,322—seventh on the list.[428]

Middlebury itself doesn't make known to the public what it pays its instructors and lecturers, but information on college data websites indicates that an average instructor at Middlebury receives $68,997[429] and an average lecturer receives $69,103[430]—both before any potential benefits. Their inclusion brings Middlebury's academic spending on environmental sustainability education to nearly $875,000, apart from the dean's salary, the salaries of affiliated faculty whose primary academic assignment is not with the environmental department, senior fellows, any support staff, teacher assistants, building costs, or other associated costs of running an academic department.

428   "2012-13 Comparative Average Salaries and Total Compensation," Middlebury College, Data About Middlebury, 2012-2013. http://www.middlebury.edu/media/view/448182/original/fac_salaries_2012.pdf

429   "Middlebury College-Instructor," Faculty Salaries, *Find the Best*. http://faculty-salaries.findthebest.com/l/12944/Middlebury-College

430   *Ibid*

Figure 17. Estimated Annual Compensation for All Environmental/Sustainability Faculty at Middlebury College

Position	Average Salary	Average Benefits	Total Average Compensation
Instructor	$68,997	n/a	$68,997
Lecturer	$69,103	n/a	$69,103
Lecturer	$69,103	n/a	$69,103
Assistant Professor	$78,775	$22,634	$101,409
Assistant Professor	$78,775	$22,634	$101,409
Associate Professor	$90,888	$28,541	$119,429
Full Professor	$129,315	$43,007	$172,322
Full Professor	$129,315	$43,007	$172,322
Total		$338,029	$119,776

Add to that the cost of Middlebury's sustainability student interns (four of them during summer 2013[431]), farm interns (four during the summer, two during each semester[432]) and five or six residential sustainability coordinators per each of the five "commons," the living-learning communities to which each student is assigned for the duration of his time at the college.[433]

Middlebury classifies and pays these student workers by the skill level required for the job (general, skilled, specialist) and by academic standing.

Figure 18. 2014 Student Hourly Wage Scale at Middlebury College[434]

Category	1st Year	2nd Year	3rd Year	4th Year
Level A General	$ 8.75	$ 8.95	$ 9.15	$ 9.35
Level B Skilled	$ 9.35	$ 9.55	$ 9.75	$ 9.95
Level C Specialist	$ 9.95	$ 10.15	$10.35	$10.55

Students are also eligible for merit raises if they return to the same job for additional years. They are not

---

431   "Meet Middlebury Summer Interns," Middlebury College, July 11, 2013. http://www.middlebury.edu/sustainability/news-events/news/2013/node/452517

432   "About Us," Organic Farm, Middlebury College. http://www.middlebury.edu/sustainability/food/mcog/about

433   "Getting to Know the Residential Sustainability Coordinators," *The Middlebury Campus*, October 13, 2010. http://middleburycampus.com/article/getting-to-know-the-residential-sustainability-coordinators/

434   "2014 Student Wage Scale," Middlebury College, January 1, 2014. http://www.middlebury.edu/offices/business/seo/paid/wagescale

allowed to work more than 20 hours per week.[435]

It is likely that at least one of the sustainability interns was classified as "specialist," since Middlebury has in the past tasked interns with gathering data and filling out annual greenhouse gas audits.[436] It is likely that the other 3 interns, as well as the 2 farm interns, were classified as "skilled," though it is possible that others may also have been classified as "specialist."[437] Assuming, then, that there was 1 specialist intern, 3 skilled interns, 2 skilled farm interns, and assuming on the conservative side that there were 25 residential sustainability coordinators (5, rather than 6, per each of the 5 commons) also classified as "skilled," and that each was paid, estimating conservatively, at the "Sophomore" level, and each worked an average of 12 hours per week (again, a conservative estimate) for approximately 14 weeks over the semester (excluding finals and midterms), that adds up to nearly $50,000 per semester, nearly $100,000 per academic year.

Figure 19. Estimated Annual Compensation of Sustainability Student Workers at Middlebury College

Position	Pay Level	Number at this Level	Sophomore Pay Rate	Hours per Week	Weeks per Semester	Total Pay per Semester	Total Pay per Year
Sustainability Intern	Specialist	1	$10.15	12	14	$1,705	$3,410
Sustainability Intern	Skilled	3	$9.55	12	14	$4,813	$9,626
Farm Intern	Skilled	2	$9.55	12	14	$3,209	$6,418
Residential Sustainability Coordinator	Skilled	25	$9.55	12	14	$40,110	$80,220
Total						$49,837	$99,674

In all, counting sustainability staff and administrators, faculty, student workers and interns, and the biomass gasification plant staff, this brings Middlebury's annual sustainability-related salary costs, as far as we can

---

435   "Working on Campus," Middlebury College. http://www.middlebury.edu/offices/business/seo/workingoncampus

436   "Each summer a student intern is hired to gather the data for the inventory, enter it, and provide an initial quality/accuracy check. This draft is then reviewed by the Sustainability Integration Office and any questions, errors, or omissions are addressed before sending a final draft for review and acceptance." See "GHG Report for Middlebury College," American College and University Presidents' Climate Commitment, May 14, 2014. http://rs.acupcc.org/ghg/3077/

437   Middlebury defines "skilled" jobs as those that "require a higher level of responsibility and some previous training or experience. Positions offer extensive on-the-job training or require certification. Students may be responsible for an aspect of a program, and/or supervising other students. Examples of Skilled Level positions include: Web assistants, Lab assistants, Lifeguards, Research assistants, Tutors, Teaching assistants, Athletic Trainers, Graders, Library or Office Associates." See "2014 Student Wage Scale," Middlebury College, January 1, 2014.   http://www.middlebury.edu/offices/business/seo/paid/wagescale

estimate, to more than $2.5 million.

Figure 20. Total Estimated Annual Compensation of Sustainability Staff and Faculty at Middlebury College

Position	Number of Staff	Estimated Salaries	Estimated Benefits	Total Compensation
Dean of Environmental Affairs	1	$197,475	$28,739	$226,214
Director of Sustainability Integration Office	1	$107,349	$21,222	$128,571
Sustainability Communications/ Outreach Coordinator	1	$41,862	$16,656	$58,518
Assistant Director of the Franklin Environmental Center	1	$61,357	$17,977	$79,334
Food and Farm Educator	1	$61,357	$17,977	$79,334
Environmental Health and Safety Coordinator	1	$76,661	$19,354	$96,015
Director of the School of Environment	1	$118,670	$22,241	$140,911
School of Environment Staff	1	$39,867	$16,909	$56,776
School of Environment Language School Per Budget Coordinator	1	$51,161	$17,492	$68,653
Coordinator for Community-Based Environmental	1	$61,357	$17,977	$79,334
Ecologist	1	$76,661	$18,588	$95,249
Instructor	1	$68,997	n/a	$68,997
Lecturer	2	$138,206	n/a	$138,206
Assistant Professor	2	$157,550	$45,268	$202,818
Associate Professor	1	$90,888	$28,541	$119,429
Full Professor	2	$258,630	$86,014	$344,644

NAS

Position	Number of Staff	Estimated Salaries	Estimated Benefits	Total Compensation
Sustainability Intern (specialist)	1	$3,410	n/a	$3,410
Sustainability Intern (skilled)	3	$9,626	n/a	$9,626
Farm Intern	2	$6,418	n/a	$6,418
Residential Sustainability Coordinator	25	$80,220	n/a	$80,220
Manager of the Central Heating Plant	1	$86,857	$19,840	$106,697
Heating Plant Operator	6	$251,172	$99,936	$351,108
Total		$2,045,751	$494,731	$2,540,482

Adding it Up

In all, Middlebury has spent millions of dollars on sustainability. According to its most recent Progress Report on file with the Presidents' Climate Commitment, Middlebury estimates that it has dispensed a total of $10-$20 million for its sustainability endeavors.[438]

The Presidents' Climate Commitment rubric operates in broad $10-million ranges. Clearly Middlebury's numbers are north of the $10 million base level, since it spent nearly $12 million on the biomass plant, but how far north is kept guarded. We undertook the task of estimating more closely what sustainability costs Middlebury.

We already estimated that the biomass plant staff cost about $457,000 per year, that maintaining the biomass plant costs approximately $50,000 (by the manufacturer's guideline of 2 percent of the cost of infrastructure), and that wood chips to feed the plant cost about $800,000 (though Middlebury calculates saving $840,000 compared to fuel costs). And sustainability staff, professors, and researchers cost about $2.5 million each year in salaries and benefits.

Other costs include keeping the environmentalist, pro-divestment Bill McKibben on staff as a distinguished scholar in residence. McKibben has occasionally taught classes at Middlebury since 2001, and when he founded 350.org in 2007, he did so with four Middlebury students. In 2010 he was announced as the Schumann Distinguished Scholar at Middlebury,[439] his position named for the Schumann Center

---

438   "Progress Report for Middlebury College," ACUPCC Reporting System, January 6, 2012. http://rs.acupcc.org/progress/390/

439   "Author and Environmentalist Bill McKibben Appointed Schumann Distinguished Scholar at Middlebury College," Middlebury College, November 9, 2010. http://www.middlebury.edu/newsroom/archive/2010/node/269059

NAS

for Media and Democracy, which is a major contributor to 350.org and to 350.org's predecessor, the 1Sky Network.[440] In 2010, the year McKibben was appointed, the Schumann center made two gifts to Middlebury: one for $1 million, and another for $200,000, according to tax documents submitted to the IRS. It did not make any further contributions to Middlebury in the years since, though it did award $100,000 to 1Sky and $211,300 to 350.org in 2011. It is likely that the Center's $1 million gift established an endowment for McKibben's position at Middlebury, and that the $200,000 constituted McKibben's salary during his first year in this new position. The return on a $1 million endowment, estimated using the 5-year average return from the S&P 500 of 15.90 percent, is $159,000.[441] This means that if McKibben continues to draw $200,000, then Middlebury is picking up a tab of $41,000. (The college was also, prior to the Schumann Center's sponsoring of McKibben, paying for McKibben's affiliation with the college since 2001, at an undisclosed amount.)

There are other costs. Middlebury prides itself on purchasing as much locally grown organic food as it can, purchasing 20 percent of its food from within Vermont State.[442] In all, it spends 32 percent[443] of a $4.2 million annual dining hall food budget,[444] for a total of $1,344,000 each year, on organic, local food.

A host of smaller programs add up to more bills, as well. Middlebury operates a campus composting facility that annually churns out 300 tons of fertilizer.[445] Every day collection trucks gather waste (about 10,000 pounds per week) from 85 campus collections bins, mix it with three parts wood chips (the same kind that go into the biomass boiler) and one part horse manure from a local horse farm, and let it ferment before spreading it on the soil around campus landscaping.[446] Middlebury does not reveal the costs of this program, but a case study from Harvard Law School's composting experiment calculated that each ton of compost cost approximately $28 above the cost of sending that ton to a landfill.[447] If Middlebury's costs are comparable, at 300 tons per year this costs $8,400.

440   Anne Journeyman, "What's Behind the Money Behind McKibben?" *Inside Philanthropy*, October 12, 2013. http://www.insidephilanthropy.com/climate-change/2013/10/12/whats-behind-the-money-behind-mckibben.html

441   S&P 500 Index SPX 5-Year Average, Quick Take Morningstar. http://quicktake.morningstar.com/index/IndexCharts.aspx?Symbol=SPX

442   "Dining at Middlebury."

443   "Local Food at Middlebury," Student Life, Dining, Local, Middlebury College. http://www.middlebury.edu/studentlife/dining/Local

444   "Dining at Middlebury," Food, Sustainability, Middlebury College. http://www.middlebury.edu/sustainability/food

445   "Composting at Middlebury," Middlebury College. http://www.middlebury.edu/offices/business/recycle/compost

446   Robert Keren, "Mining Black Gold," *Middlebury Magazine*. http://sites.middlebury.edu/middmag/2010/07/22/mining-black-gold/

447   Kate Cosgrove, "Post-Consumer Composting at Harvard Law School," Association for the Advancement of Sustainability in Higher Education, October 2011. http://www.aashe.org/files/aashe2011-materials/aashe_-_harvard_law_school_post_consumer_composting.pdf

NAS

Then there's the annual professional development conference that Middlebury sponsors, at which professors receive training on how to incorporate sustainability into their (usually non-environmental) courses. In exchange for their efforts, participating professors receive stipends and other incentives. Similar programs, such as the Piedmont Program at Emory University, pay stipends of $1,000 per professor, in addition to providing for the speakers, material, venues, and other conference costs.[448] Middlebury estimates that since its program began in 2009, 30 professors have participated, an average of 6 each year.[449] Assuming stipends are on par with Emory's, this means a cost of $6,000 per year in stipends.

Students wishing to study sustainability-related topics abroad are also eligible for $500 in sustainability study abroad fellowships. Lists of previous awardees indicate that often three students per year receive these fellowships, at a cumulative cost of $1,500 per year.[450]

Middlebury also relies on carbon offsets (550 metric tons, bought for an undisclosed price, according to Middlebury's 2014 GHG report to the Presidents' Climate Commitment) and renewable energy credits (10) that are meant to counteract some of Middlebury's remaining emissions.[451] According to 3 Degrees, an offset-selling company, 550 carbon offsets range in price from $1.75 to $12 per metric ton, for an average of $6.88 per metric ton; this indicates that Middlebury spends approximately $3,784 on carbon offsets. Ten renewable energy credits sell for a total of about $55. We estimate that Middlebury spends about $3,839 on its offsets and energy credits.

To this there are additional costs that simply are unknown. There's the bike-sharing program, the organic farm on campus, the Environmental Council Grants awarded in undisclosed amounts to student projects, the annual sustainability fair, the sustainability new student orientation, and a host of other sustainability events on campus—not to mention less tangential costs such as the cost of maintaining the office space and building used for the Office of Sustainability.

To the best of our ability, we estimate Middlebury's annual sustainability costs to be $4,920,221.

---

448   Arri Eisen and Peggy Barlett, "The Piedmont Project: Fostering Faculty Development toward Sustainability," *Journal of Environmental Education* 2006 (Vol. 38, No. 1) pp. 25-38. http://piedmont.emory.edu/documents/Articles1/Eisen26Barlett06.pdf

449   "Sustainability in the Curriculum Workshops," Sustainability, Middlebury College. http://www.middlebury.edu/sustainability/tools/courses

450   "Sustainable Study Abroad Grants," Study Abroad, Middlebury College. http://www.middlebury.edu/international/sa/sustainable/grants

451   "GHG Report for Middlebury College," ACUPCC Reporting System, May 14, 2014. http://rs.acupcc.org/ghg/3077/

Figure 21. Estimated Annual Gross Cost of Sustainability at Middlebury College

Category	Cost
Wood chips for biomass plant	$925,000
Maintenance of biomass plant	$50,000
Compensation for biomass plant staff	$457,805
Compensation for sustainability staff and faculty	$2,082,677
Bill McKibben	$41,000
Organic food for dining hall	$1,344,000
Composting	$8,400
Carbon offsets	$3,839
Professional development for faculty	$6,000
Sustainability study abroad grants	$1,500
Total	$4,920,221

These are, of course, gross costs. Even if we remove the $840,000 in fuel oil costs that Middlebury expects to save annually and the $323,709 ($250,000 inflation adjusted 2006 money) that the students estimated as the annual operations cost of running the old fuel oil plant, Middlebury's net expenditures on sustainability still total approximately $3,756,516 per year.

Figure 22. Estimated Net Cost of Sustainability at Middlebury College

Category	Cost
Estimated gross costs	$4,920,225
Fuel oil savings	-$840,000
Operation of old oil heater	-$323,709
Net costs	$3,756,516

Compared with Middlebury's complete operating budget of $292 million in fiscal year 2014, sustainability initiatives are modest.[452] They comprise about 1.2 percent of the budget. But neither are they budget-neutral (as many activists hope sustainability initiatives are), nor are they necessarily directed to the most effective, efficient ways to help the environment.

There are another 684 signatories of the American College and University Presidents' Climate Commitment who have committed themselves to similar carbon purges. Some smaller colleges, and others less advanced on their paths to carbon neutrality, surely spend less per year. But many others no doubt spend much more. Middlebury is a small college with about 2,500 students; Arizona State University, another

---

452   "General Financial Information," Middlebury College, 2014. http://www.middlebury.edu/offices/business/budget/gen_finance_info

NAS

sustainability champion, has more than 76,000, and Ohio State University has more than 57,000. These, and other large state schools, surely spend far more than Middlebury does.

Assuming, though, that Middlebury's annual $4,920,221 cost of sustainability efforts is in range of the average cost, we estimate the total annual cost of all 685 signatories endeavoring to fulfill the Presidents' Climate Commitment is nearly $3.4 billion.

There are also a host of one-time costs, such as the $11.9 million dollar biomass plant, and the additional $1.7 million that the board recently approved to modify the four non-biomass fueled heating plants to be able to run on bio-methane gas, natural gas, No. 2 fuel oil, biodiesel fuel and other types of renewable fuel rather than the more carbon-intense No. 6 fuel oil.[453] Middlebury also operates a revolving loan fund with $300,000 set aside to be used for cost-effective sustainability projects, with the goal of growing this fund to $1 million.[454] This comes to $13,900,000.

Figure 23. Estimated Fixed Costs of Sustainability at Middlebury College

Category	Cost
Biomass Gasification Plant	$11,900,000
Heating plant upgrade	$1,700,000
Green revolving loan fund	$300,000
Total	$13,900,000

Granted, Middlebury estimates in its report to the Presidents' Climate Commitment a projected $20-$30 million return on these investments, banking on "the amount of money saved annually on fuel costs for the expected lifespan of the biomass facility plus planned and likely future energy efficiency projects."[455] Whether the biomass plant will actually save money remains to be seen.

But the fixed costs, plus a mere two years' worth of annual costs, equals $23,740,442—far above the upper bound of $20 million that Middlebury reported in spending to the Presidents' Climate Commitment. Adding a third and fourth years' operations drives the cost to $33,580,884— more than $3 million above the $30 million that

Beyond the direct costs, Middlebury also faces forgone economic opportunities.

453   "Middlebury Trustees Approve 2013-2014 Budget, Sign off on new Environmental School and Hebrew Language Institute," Middlebury College, May 17,2013.http://www.middlebury.edu/newsroom/archive/524638/node/451019

454   "Green Revolving Loan Fund," Sustainability, Middlebury College. http://www.middlebury.edu/sustainability/tools/RLF

455   "Progress Report for Middlebury College," ACUPCC Reporting System, January 6, 2012. http://rs.acupcc.org/progress/390/

NAS

Middlebury hopes its biomass plant will save. This indicates that Middlebury is disclaiming significant costs in its own cost-benefit analyses and that the return on sustainability investments is significantly lower—even negative.

And beyond the direct costs, Middlebury also faces forgone economic opportunities, sidestream effects of supplying green energy, and the impediments to human health in chopping lumber for the biomass plant, erecting windmills, and installing solar panels—costs that it never reckons at all. For instance, Middlebury could choose to sell carbon credits from its own carbon reduction by means of the biomass plant—12,500 metric tons per year—probably for a total of $86,000, using the $6.88 average from the company 3 Degrees.

Diminishing Marginal Utility

Eliminating 100 percent of all greenhouse gas emissions is a gargantuan task requiring a princely sum, and it's not clear the rewards are proportionally sized. What if that money had been invested into faculty research, student programs, scholarships, or community engagement programs? What if the student courses and homework time, staff attention, and faculty research devoted to carbon neutrality had been redirected towards more timeless pursuits? At least one Middlebury student is beginning to agree. "Too many people take a fundamentalist approach to saving the environment while ignoring the fact that all actions have costs and benefits, and, sometimes, the benefits of burning carbon may indeed outweigh the costs," wrote a student, Max Kagan in a *Middlebury Campus* op-ed in February 2014.[456] Kagan contended that Middlebury rightly recruited students from diverse geographic locations, even though transporting a global student body to Vermont was far from carbon neutral: "I happen to think that a pound of carbon spent furthering the educational mission of Middlebury College is a pound we are justified in spending. Judging from the fact that most students willfully emit thousands of pounds of carbon each year in their journeys to and from campus, it appears that nearly all my peers already agree with me."[457]

Even if Middlebury were set on spending its sustainability budget on only sustainability endeavors, chasing carbon-zero is not the most effective way to cool the planet. Middlebury's initial $10-20 million investment cut down 27,618 metric tons of carbon dioxide, down from 29,882 in 2007,[458] to a net of 2,264 in 2014.[459] (That includes 12,729 metric tons of gross emissions, minus 550 in carbon offsets,

456   Max Kagan, "An Inconvenient Truth About Carbon Neutrality," *The Middlebury Campus*, February 26, 2014. http://middleburycampus.com/article/an-inconvenient-truth-about-carbon-neutrality/

457   Middlebury avers in its policies that it is responsible only for the emissions from its own geographic property and its private operations, not from student or other third-party use. But, since 2006, it does ask students travelling internationally to consider purchasing carbon offsets to account for their transportation. "Carbon Offsets," Middlebury College, Middlebury International. http://www.middlebury.edu/international/sa/sustainable/carbon_offsets

458   "GHG Report for Middlebury College," ACUPCC Reporting System, November 26, 2008. http://rs.acupcc.org/ghg/441/

459   "GHG Report for Middlebury College," ACUPCC Reporting System, May 14, 2014. http://rs.acupcc.org/ghg/3077/

NAS

10 in Renewable Energy Certificates, and 9,905 in sequestration credits for the forests that Middlebury maintains on its property.) That's a price ranging from $362 to $724 (an average of $543) per metric ton of carbon that Middlebury refrained from emitting. If that price remains constant for the last 2,264 metric tons that Middlebury has yet to eradicate, the college will spend somewhere between another $800,000 to $1.6 million achieving carbon neutrality. More likely, though, those last emissions will be more difficult and more expensive to eliminate.

By comparison, the Environmental and Energy Studies Institute estimates that the EU under a cap and trade program launched in 2005 has spent €10 to €30, or $12 to $36, per metric ton eliminated, and Pigouvian carbon taxes in Canada reduced emissions at a cost of $15 per metric ton.[460] The most expensive program profiled in the EES study was Tokyo's, which cost $142 per metric ton.

Figure 24. Cost Per Metric Ton of Greenhouse Gas Reduction

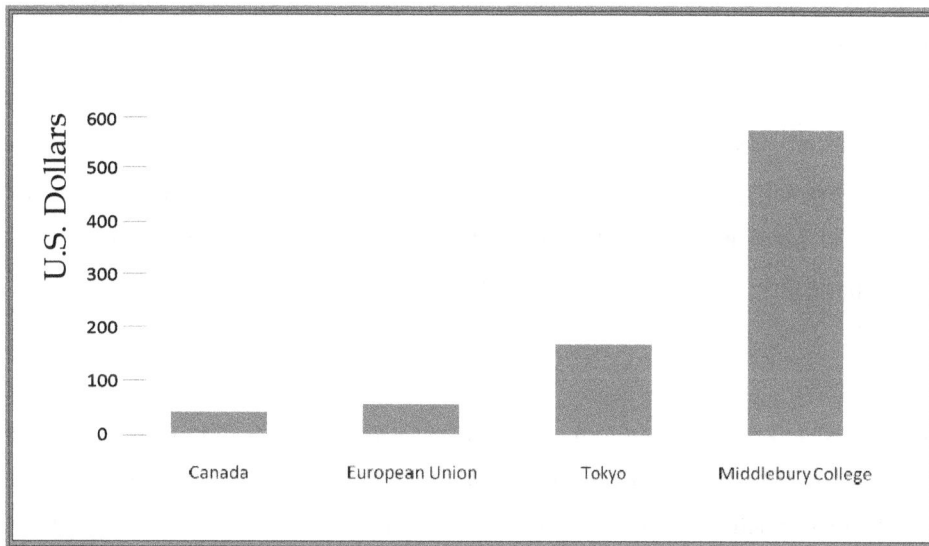

Perhaps it sounds unfashionably frugal to expect a good bang for your buck, but the reality is that Middlebury might improve the environment several hundred-fold if it invested its sustainability-slotted funds not into its own, already eco-friendly campus, but into easier, cheaper solutions elsewhere.

Middlebury's multi-million dollar effort to achieve zero-carbon is just a drop in the bucket of sustainability spending. Hundreds of other colleges, universities, nonprofits, cities, counties, and other local governments have set similar targets and spent similar amounts. Meanwhile international delegations to the September

---

460   "Carbon Pricing Around the World," Environmental and Energy Study Institute, October 2012. http://www.eesi.org/files/ FactSheet_Carbon_Pricing_101712.pdf

NAS

2014 UN climate summit in New York City recently began the process (to be concluded in Paris in 2015) of preparing carbon cuts, not quite to zero carbon, but still sizeable, pricey carbon diets with enormous repercussions on international economics.

And sustainability generates other costs across the national economy. There are enormous failed investments by the Federal government in companies such as Solyndra ($535 million), Abound Solar ($400 million), Beacon Power ($43 million), and A123 ($249 million), to name a few of the "sustainable" energy companies that went bankrupt when their product failed to provide efficient, reliable services at market price, despite millions of dollars of federal money to prop them up.[461]

EPA regulations drive up energy prices and impede industrial development. The U.S. Chamber of Commerce's Institute for 21st Century Energy found in a May 2014 study that by 2030, EPA-proposed carbon regulations could result in an annual loss of $51 billion in GDP, lower disposable household income by $586 billion, increase electricity costs by $289 billion, and put approximately 224,000 Americans out of work each year.[462]

Then there are tax incentives and federal grants for green construction. At the heart of the American Recovery and Reinvestment Act of 2009 (the "stimulus") was $5 billion set aside for green retrofits on buildings—much of which got misdirected, as in Delaware, where contractors authorized to perform simple repairs (insulating attics, sealing gaps) instead opted to entirely replace utilities and building infrastructure.[463]

Meanwhile the mandating of ethanol as a component of fuel for automobiles under the Renewal Fuel Standard (created in 2005 with the Energy Policy Act, strengthened in 2007 by the Energy Independence and Security Act) has inflated corn prices. The Congressional Budget Office found that compliance with the mandate "would increase total spending on food in 2017 by $3.5 billion."[464]

And switching from coal and oil to wind and solar is an unrealistically expensive change. Germany, the most aggressively pro-renewable energy country in the world, has reluctantly backed away from its ambitious goals to cut its emissions to 40 percent of 1990 levels by the year 2020, after the costs

461   Steve Hargreaves, "Obama's Alternative Energy Bankruptcies," *CNN*, October 22, 2012. http://money.cnn.com/2012/10/22/news/economy/obama-energy-bankruptcies/

462   *Assessing the Impact of Proposed New Carbon Regulations in the United States*, Institute for 21st Century Energy, U.S. Chamber of Commerce, 2014. http://www.energyxxi.org/epa-regs#

463   Stephen Clark, "Obama's $5 Billion Weatherizing Program Wastes Stimulus Funds, Auditors Find," *Fox News*, April 14, 2011. http://www.foxnews.com/politics/2011/04/14/obamas-5-billion-weatherizing-program-wastes-stimulus-funds-auditors/

464   "The Renewable Fuel Standard: Issues for 2014 and Beyond," Congressional Budget Office, June 2014, pg. 15 http://www.cbo.gov/sites/default/files/cbofiles/attachments/45477-Biofuels2.pdf

NAS

associated with renewable energy became prohibitive. Vice Chancellor Sigmund Gabriel announced his plans in November, remarking, "It's clear that the [2020 CO2] target is no longer viable."[465] The renewable energy transition policies, called Energiewende, phased out nuclear energy and attempted to increase solar and wind substitutes. But costs skyrocketed. German residents now pay the second highest energy rates in Europe, as well as 106 billion Euros in tax subsidies to renewable energy companies over the course of 2010-2013. Increasingly, customers and power generators turned to coal instead.[466]

The unreliable nature of renewable energy sources (the wind doesn't always blow, or the sun always shine) combined with their exorbitant costs have led some to argue in favor of fossil fuels not just as a matter of economics, but of ethics. Societies need reliable access to energy. Scholars such as Alex Epstein (*The Moral Case for Fossil Fuels*) and Robert Bryce (*Power Hungry: The Myths of "Green" Energy and the Real Fuels of the Future*) note the rising quality of healthcare, life expectancy, standards of living, and educational attainment that correspond with cheap, abundant, reliable energy. Bryce has cast coal as a "social justice" issue, writing in *National Review Online*, "Coal is an essential fuel for combating energy poverty." His research showed that

> between 1990 and 2010, about 1.7 billion people gained access to electricity. Of that number, some 830 million people gained access owing to coal. Natural gas came in second, with about 380 million, and hydro came in third with about 290 million. Put another way, over that two-decade period, for every one person who gained access owing to solar and wind energy, four gained access owing to hydro; six gained access owing to natural gas; and 13 gained access owing to coal.[467]

Middlebury students, however, fighting hard for carbon neutrality and fossil fuel divestment, have stubbornly marked out their opposition to fossil fuels as a pro-social justice position.

Sustainability as Business

As government policies have pressured institutions to submit to environmental regulations, and environmental activists have pressured other institutions to voluntarily adopt stricter standards, hundreds of businesses have set about manufacturing ways to get them from carbon-dependent to carbon-zero. We recognize the value of many protective measures and technological innovations that keep our air and water clean and preserve our natural terrains. But we also notice an accelerating movement pushing society to accept and prefer all things green, regardless of cost. In response, the economy has shifted, and not always in a healthy way. Sustainability is big business.

---

465   "Germany Plans To Withdraw From Binding 2020 Climate Targets," *Spiegel*, reprinted by the Global Warming Policy Foundation, November 16, 2014. http://www.thegwpf.com/germany-announces-withdrawal-from-binding-2020-climate-targets/

466   "Germany's Renewable Energy Transition Misses Carbon Reduction Goals," Institute for Energy Research, September 30, 2014. http://instituteforenergyresearch.org/analysis/germanys-renewable-energy-transition-misses-carbon-reduction-goals/

467   Robert Bryce, "The Social Justice of Coal," *National Review Online*, November 6, 2014. http://www.nationalreview.com/article/392167/social-justice-coal-robert-bryce

In the cleft of radical activism and regulatory activism, a new branch of industry has sprung up, feeding off of fears of tighter regulations and fears of climate change. With an inelastic demand for products and services designed for zero waste, zero carbon, and zero "impact," companies needn't compete on price, only on effectiveness at curbing perceived environmental hazards.

That isn't necessarily a strike against the eco-entrepreneurs, of course. Engineering better ways to clean up trash or responsibly extract and use resources is wise—and in large part, they're simply responding to a change in consumer tastes and preferences. As a market opens up for these new services and products, it only makes sense for enterprising businessmen to serve this niche clientele. What is more astounding, though, is consumers' ardent demand and their willingness to pay for such services.

One such industry is the new market for inventive recycling techniques. Dozens of colleges and universities partner with Terracycle, a self-billed "trash alchemist" company that finds ways to repurpose hard-to-recycle items. Terracycle specializes in "upcycling," the process of using trash to create a new product, ideally worth more than its ingredients. The idea is to use waste products whole—gluing together Oreo packaging wrappers into a kite or sewing Clif Bar wrappers into the body of a backpack—without melting, smashing, or otherwise demolishing the waste.

This trash to treasure program means a boon for Terracycle. It started in 2001 when Tom Szaky, then a Princeton freshman, got the idea of feeding leftovers from the Princeton cafeteria to worms, collecting their excrement, and selling it in used soda bottles as fertilizer. The business grew to such success that Szaky left school to run his business, but after a 2007 altercation with Scotts Miracle-Gro, he switched gears, giving up the fertilizer business and repurposing (perhaps upcycling?) his branding into a recycling company with an entrepreneurial twist.

Unlike some recycling organizations spurred by philanthropy, Terracycle is driven by profit. Szaky found a market niche catering to the needs of the environmentally-conscientious who felt guilty throwing things away. Many cities already provided recycling services for certain metal, plastic, paper, and glass items. But Terracycle came up with ways to repurpose hard- to- recycle items such as Cheetos bags and Capri Sun pouches by cleaning the wrappers, unwrapping them, laying them flat, and sewing them back together. Szaky started "brigades" for food wrappers, plastic gloves, bottle corks, even cigarette butts. From these sprang the Oreo package kites and Clif Bar wrapper

backpacks—along with coasters made from computer circuit boards, Doritos pencil boxes, Lays potato chip messenger bags, and Sun Chips placemats. Terracycle also offers more conventional recycling options: plastic fence posts, benches, spray bottles, flower pots, and (ironically) trash cans made from melted plastic and other waste.

Terracycle is a godsend for institutions such as Middlebury intent on reducing their waste-lines. They can send to Terracycle their Oreo packaging and Clif Bar wrappers and soothe their consciences with the knowledge that the plastic, once opened, laid flat, and sewn together, will turn into a lightweight plastic kite or a child's backpack—no trash involved. Middlebury's relief at finding a way to clear its trash conscience is evident in the name it bestowed upon its Terracycle partnership: the plastic tubs where students can deposit their upcycling-bound trash are called "Terracycle Sin Bins."[468]

These trash indulgences prove economically inefficient. Terracycle's business plan relies on what Szaky calls "sponsored waste," wherein institutions pay for the privilege of declaring their products "zero waste." It's free market economics meets environmental zealotry at its worst. Product brands pay Terracycle engineers to find ways to repurpose the unreusable parts of their products, typically the packaging. The brands then advertise themselves as especially eco-friendly, and benefit from the patronage of zero-wasters like Middlebury College, who can save up their wrappers and mail them to Terracycle, in exchange for mini donations that Terracycle gives to various charities they choose from. Terracycle then manufactures upcycled products and sells them as eco-friendly alternatives to mainline carbon- based department store offerings. It's a clever intrusion into the free market. Terracycle benefits from free inputs (trash) and monetary sponsorships from brand names. The zero-wasters earn a claim to being diligent recyclers and a feeling of easy philanthropy. And the purchasers of the upcycled products get to feel good about their environmentally-thoughtful purchases and get to wear their commitments, if not on their sleeves, at least on their handbags and backpacks.

There are big opportunities for those who would profit from others' carbon phobias. McKenzie Funk, a journalist from Seattle, chronicles some opportunistic climate change entrepreneurs in his book *Windfall: The Booming Business of Global Warming* (Penguin Press, 2014). Funk spent six years traveling the globe to document Israeli desalinization plants—originally invented to freeze, melt out the salt,

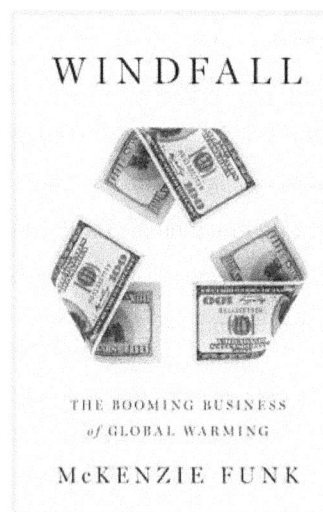

WINDFALL

THE BOOMING BUSINESS
of GLOBAL WARMING

McKENZIE FUNK

---

468 "Terracycle Sin Bins," Environmental Council Grants 2011-2012, Middlebury College. http://www.middlebury.edu/sustainability/fech/ec/grants/2011-12

NAS

and make water from the Dead Sea drinkable—doubling as fake snow producers for melting ski slopes in the Alps, Canadian troops defending the Northwest Passage (thawing and passable for foreign trade for the first time in centuries), and Africa's proposed Great Green Wall, a 4,700-mile-long, 10-mile-wide strip of trees meant to hold back the expansion of the Sahara Desert.

Some of the institutions Funk profiles fill real needs that existed before climate change loomed large in the public mind—the Great Green Wall was first proposed in 1952 by Richard St. Barbe Baker, an English forester who saw the tree-band as a way to reduce land mismanagement, over-grazing, and over-harvesting. Funk also spends time in the Netherlands, where Dutch engineers trained in building dikes for the lowland country design sea walls for flood zones in island nations.

But others prey off people's fears: private firefighting squads in Los Angeles for expensive, high-insurance homes, or investment companies constructing "climate change funds" that invest in windmills and solar panels and all the other markets expected to rise in value once climate change hits: biotech plants re-engineering agriculture for warmer temperatures, dredging rigs that rebuild flooded islands, reinsurance companies expected to jack up insurance rates, even grocery stores presumed to profit from thepredicted food scarcity. These funds bank not only on what they expect to become intrinsically valuable, but on what quavering markets will begin to prize as they come to fear global warming and its accompanying regulations. "There's another possible response to melting ice caps and rising sea levels" besides regulating and nudging consumers towards carbon cuts, Funk writes, "a response that is tribal, primal, profit-driven, short-term, and not at all idealistic. Every man for himself. Every business for itself."[469]

Saving Carbon

What exactly does sustainability cost? No one knows exactly, and that silence is telling. Cost-benefit analyses seem to have gone with the wind. Perhaps they were blown there by precautionary principle-fanned sustainability-driven wind turbines. The extravagant costs of sustainability paired with the economic inefficiencies suggest that as much as colleges like to talk about saving money through high-efficiency light bulbs and Energy STAR-certified appliances—or, in the case of Middlebury, its supposedly cost-neutral Biomass Gasification Plant—the real goal is to save carbon. Colleges' eager assent to expensive reforms in a time of shrinking budgets indicates the power of the sustainability movement on college campuses.

469   McKenzie Funk, *Windfall: The Booming Business of Global Warming*, New York: Penguin Press, 2014, pg. 8.

NAS

CHAPTER 6: DIVESTMENT: SUSTAINABILITY'S LAST FRONTIER

The agenda for the May 4, 2013, meeting of the Swarthmore College board of managers was laid out clearly: a PowerPoint presentation from the chairman of the investment committee, short commentsfrom several students, and an open session for remarks from interested professors, staff, and students. The board had convened in Swarthmore's Science Center auditorium at 11:00 AM in an open meeting to consider divesting the college's endowment from its stock in fossil-fuel based energy companies, and in due time, it would hear from members of the college community on both sides of the debate.

Students had been agitating in favor of divestment for several years, and the administration had met with many of the pro-divestment students before—upwards of 25 times, according to Kate Aronoff, one of the leaders of the campus's student divestment organization, Mountain Justice.[470] The group took its name from a class trip over fall break in 2010 to West Virginia coal country, where visiting professor of peaceand conflict studies George Lakey led his students to meet the environmentalist Larry Gibson ("Keeper of the Mountains") and to witness the coal mining industry's mountaintop removal operations. "If you don't do anything about this, then I've wasted my time," Aronoff recalls Gibson told the students.[471] They returned to campus and launched Swarthmore Mountain Justice in October 2010. In "solidarity" with the West Virginians who'd resisted the mining industry, Mountain Justice began pressuring Swarthmore to give up its profitable holdings in coal, as well as in other fossil fuel companies.

"The difference between the majesty of the Appalachian Mountains that hadn't been ruined and the mountain peaks that were flattened was so stark," Aronoff told the *New York Times* for a front-page Business Day article on Swarthmore's divestment efforts in May 2014, one year after the board of managers meeting.[472]   "We saw and heard about how toxic the coal mining industry is and how much of the economy is structured around coal mining. It was a moment when the connection between economic injustice and environmental injustice was just so clear."

At the May 4th meeting, the board had invited two representatives of Mountain Justice to sit on a panel with them. Swarthmore President Rebecca Chopp and a "Quaker moderator" (a nod to Swarthmore's religious heritage) presided. As it happened, there was more quaking than peacemaking—and not quaking of the religious kind. Before the chairman of the Board Finance Committee, Chris Niemczewski, had gotten through two slides in his presentation, the Mountain Justice students began talking over him,

470   James B. Stewart, "A Clash of Ideals and Investments at Swarthmore," *New York Times*, May 16, 2014. http://www.nytimes.com/2014/05/17/business/a-clash-of-ideals-and-investments-at-swarthmore.html?gwh=6414F82B4F6E7B2A8FF8BE9189A8BE7A&gwt=pay&assetType=nyt_now&_r=0

471   Kate Aronoff, interview with Rachelle Peterson, October 29, 2014.

472   Stewart, "A Clash of Ideals and Investments at Swarthmore."

seizing the microphone and recounting the "frustration, anger, and hurting on our campus" and the "daily acts of aggression" that dirtied Swarthmore's social environment. One of the students, Pat Walsh, warned that he was "creating a platform for student voices" and turning the tables on the "authority" of the board, and that anyone who left the meeting would evince his or her "unwillingness to listen and to respond to the needs of this community." He signaled a cohort of more than 100 students who had waited outside and then flooded the room, placards and posters in hand.[473] "Check ur ignorance," and "This is social responsibility," the signs read.

Grabbing the microphones from the board members, the students began a series of tirades about the need for "radical, emancipatory change" at Swarthmore. Two students from the student group Swatties for a DREAM asked Swarthmore to support undocumented students and to lobby Congress to grant them citizenship.[474] A young woman from the Sexual Misconduct Advisory Resource Team (SMART) patiently explained the need for sensitivity training for all administrators, and especially for professors, because the faculty, with their "privileged power positions" over their classes, "can be very damaging to the students."[475] Another complained about the lack of diversity on campus—only 93 black students—and how the college treated minorities as "diversity tools" to advertise the campus's racial heterogeneity, "but then you don't take care of us. You don't treat us as individuals."[476] Hope Brinn, a sophomore suing the school for "blatant, blatant noncompliance with the federal law" for mishandling her reports of stalking and sexual harassment, accosted the board: "Serial rapists—why are they still on this campus? We are fully aware of them."[477] Swarthmore has since been embroiled in an investigation with the Department of Education's Office for Civil Rights; it also settled a case with a male student who was expelled for sexual assault after the College mishandled his case and was found to lack an impartial judiciary procedure.

> *"Check ur ignorance," and "This is social responsibility," read the signs held by pro-divestment student protestors.*

---

473   "Open Meeting with the Board of Managers," Swarthmore Mountain Justice, YouTube, May 4, 2013. https://www.youtube.com/watch?v=00Med0treVE

474   "Student Testimony at Swarthmore Board Meeting – Swatties for a DREAM," SwarthmoreInterACTS, YouTube, May 9, 2013. https://www.youtube.com/watch?v=lDVb-Gr-MFI&list=UUh7suaodj9Vppl00hGIqkZQ

475   "Student Testimony at Swarthmore Board Meeting – Miriam Hauser," SwarthmoreInterACTS, YouTube, May 9, 2013. https://www.youtube.com/watch?v=-3iix9cqgHg&index=8&list=UUh7suaodj9Vppl00hGIqkZQ

476   "Student Testimony at Swarthmore Board Meeting: Jusselia Molina," SwarthmoreInterACTS, YouTube, May 11, 2013. https://www.youtube.com/watch?v=QalSXlMQ1Rg&list=UUh7suaodj9Vppl00hGIqkZQ

477   "Student Testimony at Swarthmore Board Meeting: Hope Brinn," SwarthmoreInterACTS, YouTube, May 11, 2013. https://www.youtube.com/watch?v=0Qv6JHm8YEM&index=3&list=UUh7suaodj9Vppl00hGIqkZQ

NAS

Eventually Sara Blazevic and Nathan Graf presented Mountain Justice's divestment proposal and laid out a timeline of action that they expected the board to meet.[478] They wanted the board to read and begin implementing Mountain Justice's divestment plan (total divestment from 16 fossil fuel companies) by August 2013, if necessary calling "an ad hoc meeting, whenever, in the summer"; publish a report of the board's plans by September 1st; release an official decision from the board at the September 27-28 board meeting; and by October 6-7 demonstrate in detail Swarthmore's full implementation of the divestment plan. "For me, this is a question of accountability," Blazevic explained. "What we're doing right now in this room is holding you, the board of managers, and the administration of Swarthmore College, holding you accountable. We want everyone in this room to hold each other accountable."[479] The board members and the president of the college sat quietly in their chairs as though nothing had happened.

In all, 28 students confronted the board with their grievances, stretching what was to be a 1-hour meeting to an hour and a half. When one elderly voice called out from the audience politely asking for the chairman to be allowed to finish his presentation and to return to the agenda, one of the students responded, "He can get in line if he'd like to make a statement!"[480] Another board member, Dulany Bennett, a conciliatory silver-haired woman, stepped up to the microphone to concede, "I think the board of managers are prepared to stay, just to sit and listen to what you have to say."[481] At the end of the students' demands, two board members spoke. Nate Erskine confessed,

> It's obvious that we do not have the proper trust and communication in the Board of Managers and the students. I care for you guys, I want you guys to feel safe and empowered and that Swarthmore is doing all that it can for you.[482]

Susan Levine, the other board member who responded, assured the students, "All the issues that you have been raising today we have been talking about with great seriousness and concern in our Board meetings."[483]

Clapped Down

All of this was captured on video uploaded to YouTube by Mountain Justice as a tribute to their coup.[484]

---

478   "Student Testimony at Swarthmore Board Meeting: Swarthmore Mountain Justice," SwarthmoreInterACTS, YouTube, May 11, 2013. https://www.youtube.com/watch?v=dq6jMIeNRjY

479   *Ibid.*

480   "Student Testimony at Swarthmore Board Meeting: Ian Perkins-Taylor," SwarthmoreInterACTS, YouTube, May 11, 2013. https://www.youtube.com/watch?v=19encr_eB4s&list=UUh7suaodj9Vppl00hGIqkZQ&index=4

481   *Ibid.*

482   Andrew Karas and Cristina Abellan-Matamoros, "Student Protesters Take Over Open Board Meeting, State Wide Array of Concerns," *Swarthmore Daily Gazette*, May 6, 2013. http://daily.swarthmore.edu/2013/05/06/student-protestors-take-over-open-board-meeting-state-wide-array-of-concerns/

483   *Ibid.*

484   "Student Testimony at Swarthmore Board Meeting: Watufani M. Poe," SwarthmoreInterACTS, Youtube, May 9, 2013. https://

Ten minutes into the tirade, Danielle Charette and Preston Cooper, two students from the Swarthmore Conservative Society opposed to divestment, stood up from their seats in the audience to express their disapproval of the mob's takeover. Cooper, an economics major in the honors college, told us that he had planned to deliver a few remarks during the open comment section about how divestment would harm the college endowment and make no recognizable dint in fossil fuel companies' profits. But the protestors cancelled the comment section and instead required that anyone—including the board members—who wanted to speak must stand at the back of their line, now extending out the back doors.

Charette tried to remonstrate with the protestors that "The board has given us an hour of their time" but that the students had "hijacked the meeting." In response, one of the protesters shouted, "Get in line!" and another called, "You are also free to leave."[485] The others began clapping in unison, clapping her down and drowning out her voice.

A Mountain Justice student leader, Nathan Graf, acknowledged to us over the phone that clapping down Charette was "not appropriate—not okay." Instead, Mountain Justice leaders had discussed using the clap-down technique only at "appropriate" times: that is, not against their fellow students but against any "board member who tried to filibuster."[486]

Charette and Cooper approached the moderator, who did nothing; President Rebecca Chopp, who acknowledged that the behavior was "outrageous" but refused to take action; and the Dean of Students, Liz Braun, who also offered no response. Meanwhile, angry Swarthmore student Watufani Poe stepped up to the microphone to fulminate about an incident in which an intoxicated student urinated on the Intercultural Center, an act he took personally as anti-gay bigotry.

Danielle Charette (in green) asks President Chopp to restore the Swarthmore board meeting to order.

That semester at Swarthmore had been turbulent, and Mountain Justice set itself up as a sponge for all campus grievances, taking on "justice" causes not just for the mountains, but for gender equality, gay issues, diversity, class egalitarianism, and labor, as evidenced by the speakers they had lined up for the meeting. "Mountain Justice began to speak for the whole campus," Charette told us, as the group began

---

www.youtube.com/watch?feature=player_embedded&v=TS3Xa9UMZu8#.

485   *Ibid.*

486   Nathan Graf, interview with Rachelle Peterson, September 23,   2014.

NAS

allying with other campus activist groups.[487]

Mountain Justice's blog elaborates on the link between the environment and sustainability's other two circles, society and the economy: "Climate change is not just an environmental issue — it is a concern for anybody interested in confronting racism, sexism and economic injustice."[488] In addition to campaigning for divestment, the group also works to promote awareness of "how women are severely impacted by climate change," the way that "racism and environmental destruction intersected in the aftermath of Hurricane Katrina," and "how climate change disproportionately impacts poor communities in the Global South."[489]

After Poe completed his rant, accompanied by the cheers of the protesters, he ran up the auditorium aisle and out of the room. Two other Mountain Justice students then stepped forward to tell Charette to get in line or leave.

Afterwards, Charette wrote an op-ed for the *Wall Street Journal*, "My Top-Notch Illiberal Arts Education," wryly deploring the unraveling of civility and freedom of speech at her college. She noted that more than a week after the incident, "No administrator has condemned the takeover of the board meeting," and asked, "If that tantrum doesn't qualify as disorderly conduct and outright intimidation, what does?"[490] In contrast George Lakey, the peace and conflict studies professor who first encouraged his students to start the divestment campaign, praised the meeting as a "useful" affront to the "self-absorption" of students who were insensitive to their "vulnerable" fellow students.[491] He also saw the takeover and clap-down as a needed wake-up call to the "conflict-averse 1-percent" who had no idea how to respectfully interact with their "oppressed" peers.[492]

Two days before Charette's op-ed, President Chopp had explained to Stanley Kurtz (covering the incident at *National Review Online*) that she had exercised reticence out of deference to the college's "Quaker roots of tolerance." She rationalized that

at Swarthmore we try to allow students to have their say, and we seek to listen to their points of view

---

487 Danielle Charette, interview with Rachelle Peterson, September 8, 2014.

488 "What is Climate Justice?" Swarthmore Mountain Justice. http://swatmj.org/ourcampaign/#climatejustice

489 *Ibid.*

490 Danielle Charette, "My Top-Notch Illiberal Arts Education," *Wall Street Journal*, May 15, 2013. http://online.wsj.com/news/articles/SB10001424127887324216004578483080076663720

491 George Lakey, "Op-Ed: Overlooked Aspects of the Student Intervention in the May 4th Board of Managers Meeting," *Swarthmore Daily Gazette*, May 14, 2013. http://daily.swarthmore.edu/2013/05/14/op-ed-overlooked-aspects-of-the-student-intervention-in-the-may-4th-board-of-managers-meeting/

492 George Lakey, "Swarthmore College's Rude Awakening to Oppression in Its Midst," *Waging Non-Violence*, May 21, 2013. http://wagingnonviolence.org/feature/swarthmore-colleges-rude-awakening-to-oppression-in-its-midst/

before we make decisions. Sometimes this is difficult and messy, sometimes people do not agree, and sometimes it does not work the first   time.[493]

A year later, in June 2014, Chopp resigned from Swarthmore, the #3 ranked liberal arts college in the country, according to *U.S. News and World Report*, to become chancellor of the University of Denver, #91 among national universities. In her farewell letter, Chopp acknowledged that the decision process had been "one of the most difficult ones of my professional life." She said she was driven by "both personal and professional desires," including her husband's health challenges and the proximity of family in Denver.[494]

Whatever Chopp's intents about reticence, free speech, and trying to "allow students to have their say," Mountain Justice, for its part, was uninterested in letting everyone have his or her turn to speak. Shortly after the takeover, Aronoff posted an article on Swatoverlaps, a student blog, titled, "'What Swarthmore Really Stands For,' or F*** Your Constructive Dialogue."[495] Only in Aronoff's piece, there were no asterisks. There, Aronoff criticized those who "come to the defense of civilized debate" and argued that the forums of rational debate are themselves too elite and hence favorable to the oppressor rather than the oppressed. Mere tolerance is too passé. Radical action is instead required. Hence the angry mob on May 4th.

Scheduled Crises

In September 2013, Swarthmore's board of managers officially dismissed the petition to divest. In an open letter to the Swarthmore community, chairman Gil Kemp wrote that the board's decision against divestment was "broad and deep" and arrived at from "several different paths." He cited the cost, the unclear value of a "symbolic act" against the fossil fuel industry, and the importance of alternative measures—such as hiring the college's first sustainability director and encouraging interdisciplinary study of environmental issues.[496]

But Swarthmore Mountain Justice has not backed down. If anything, in the fossil fuel divestment movement, defeat seems to be a badge of honor. At a recent divestment rally in New York City, students identified themselves with different characteristics that a central leader called out, such as LGBTQ or first-generation college student. When the leader—incidentally, Kate Aronoff, now a Swarthmore graduate

493   Stanley Kurtz, "Swarthmore's President Chopp Replies to my Queries," *National Review Online*, May 13, 2013. http://www.nationalreview.com/corner/348138/swarthmores-president-chopp-replies-my-queries-stanley-kurtz

494   Rebecca Chopp, "Community Message," President's Office, Swarthmore College, June 12, 2014. http://www.swarthmore.edu/presidents-office/community-message-president-rebecca-chopp

495   Kate Aronoff, "'What Swarthmore Really Stands For,' or, F*** Your Constructive Dialogue," Swatoverlaps, May 2013. http://swatoverlaps.tumblr.com/post/49900407586/what-swarthmore-really-stands-for-or-fuck-your

496   Gil Kemp, "An Open Letter on Divestment," Swarthmore College Board of Managers, September 11, 2013. http://www.swarthmore.edu/board-managers/open-letter-divestment

NAS

volunteering with the Divestment Student Network—called out "anyone who's gotten a 'no!' on their divestment campaign," the circle of about 75 students erupted in cheers for those who stepped forward.

Divestment campaigners are organized, angry, and determined to win. At a recent rally with the national Divestment Student Network they chanted with gusto, "I believe that we will win! I believe that we will win!" They have ample resources to help them. 350.org staff serve as divestment campaign advisers, as do the Responsible Endowments Coalition staffers. The student-run Divestment Student Network holds regional training sessions and connects students across campuses. In spring 2014, students from 75 campuses forged a plan to unify their dispersed campaigns in a 2014-2015 "Power-Building Initiative." They divided the country into regions and sub-regions, appointed student regional directors, and developed schedules of phone call check-ins to keep campaigns on track.

Figure 25. Divestment Student Network Roles

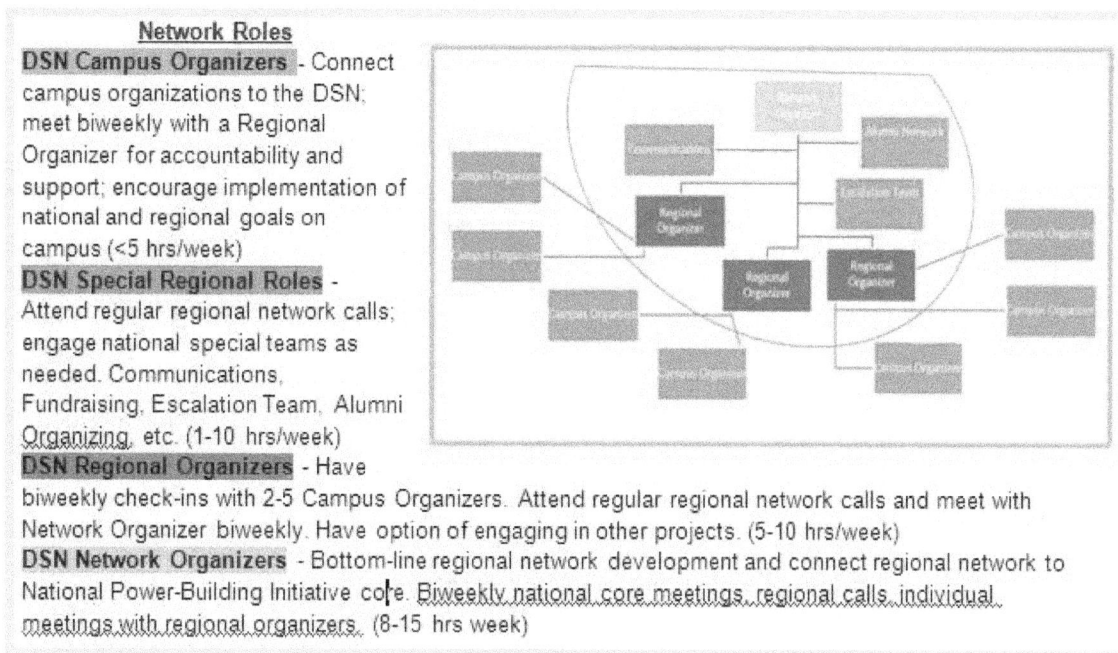

Swarthmore Mountain Justice lead organizers Sara Blazevic and Guido Girgenti, along with their mentor Katie McChesney from 350.org, held a training session in New York City on September 20, 2014 the day before the People's Climate March. Mountain Justice addressed student organizers from around the country who were interested in "escalating" their campaigns until their boards of trustees capitulated. Blazevic and Girgenti distributed copies of their action schedule for the entire school year, broken down month by month into specific action items and target goals that aim, in the end, for "kicking a**"—the board's, presumably. (The schedule is included below.)

NAS

	September	October	November Fall 2014	December	Jan
Key dates		Fall break: October 12-20			Classes begin: 19th
Goals			Maintain core. Shift campus majority into active allies through act-recruit-train cycle: return to ladder of engagement for a second or third time) & passive allies (wear orange squares, support divestment if asked, maybe engaged once). Win high-profile alumni support. Prepare student, alumni, parent base for NVDA. Coordinate with other campuses.	Engage with Board members following semester of kicking ass --> get to place where we can ask them in early February to give soft endorsements. Invite them to publish in Phoenix: stand apart from monolith of the Board, express openness to divestment.	
Projects	Mass training with everyone from PCM & their roommate week after PCM. Theory of change & strategy. Follow-up ask for orange square distribution. Need to be prepared when story of fossil fuel stock dropping (post-PCM) hits, drop a hard-hitting large media campaign, including large Board in Parrish (people who are for vs. against divestment). End of month: revisit departmental organizing.	Act-recruit-train post PCM into fall break. Marcie speaking post-fall break as thing for us to plan for through fall break: galvanize energy for another mobilization push, etc. Could get econ dept to cross-endorse this event as academic event??	Dept. organizing for September: re-assess what other outreach tactics. Organize POLS, ENVS, Soc/Anth, PEAC, Math, Bio. Mass interest meetings: create clear ways for supporters to publicly affiliate w/ campaign. Public media campaign: posters & infographics across campus showcasing all endorsements of movement (pastor in PA?). Develop alumni core: alumni withholding donations. Alumni working on 1) fund 2) letters. NationBuilder is huge for alumni. Have sign-up for network of alums --> alumni petition in support of divestment. Get up on NationBuilder before PCM. 100 faculty pledged by mid-March --> 60 faculty by end of November.		Ask of 3 members of Board to announce that they will not stand in the way of divestment (makes room for media in early February). David Gelber: creating space for him to be a climate movement hero.
Targets (groups being moved)			All campus allies & neutral's: libera's. ICC/BC, WRC, StuCo: greens. Natural sciences & Econ. High-profile alums & donors.		
Actions & events	Pre-PCM: week after retreat, open meeting to turn a lot of people out: intro campaign, etc. 3 orange square ask around: come to PCM & come to this meeting next week to prepare for PCM. Post-PCM: mass training. Media & comms plan for alumni & faculty to blast out endorsements.	Marcie Smith lecture Oct. 21 or 22.	1st week: low barrier, highly visual. BoM 1: Mark Wallace & student give positive vision & narrative. YouTubed. BoM 2: Reiterate positive vision, make direct ask to ON. Public debate between M.J. fossil fuel industry supporter and Swat Admin. Activities Fair.		
Metrics			Train & retain 150 people over course of semester. 75 do more leadership development by end of the fall. 60 faculty pledge. 1,000 alumni e-mails. 10-12 people in the core.		

NAS

Figure 26. Swarthmore Mountain Justice 2014-2015 Schedule

	September	October	November	Fall 2014	December	Jan
Key dates		Fall break: October 12-20			Engage with Board members following semester of kicking ass --> get to place where we can ask them in early February to give soft endorsements. Invite them to publish in Phoenix. stand apart from monolith of the Board. express openness to divestment.	Classes begin: 19th
Goals	Mass training with everyone from PCM & their roommate week after PCM. Theory of change & strategy. Follow-up ask for orange square distribution. Need to be prepared when story of fossil fuel stock dropping (post-PCM) hits. drop a hard-hitting large media campaign. including large Board in Parrish (people who are for vs. against divestment). End of month- revisit departmental organizing.	Act-recruit-train post PCM into fall break. Marcie speaking post-fall break as thing for us to plan for through fall break. galvanize energy for another mobilization push. etc. Could get econ dept to cross-endorse this event as academic event??		Maintain core. Shift campus majority into active allies through act-recruit-train cycle. return to ladder of engagement for a second or third time) & passive allies (wear orange squares. support divestment if asked. maybe engaged once). Win high-profile alumni support. Prepare student. alumni. parent base for NVDA. Coordinate with other campuses.		Ask of 3 members of Board to announce that they will not stand in the way of divestment (makes room for media in early February). David Gelber: creating space for him to be a climate movement hero.
Projects				Dept: organizing for September: re-assess what other outreach tactics. Organize POLS. ENVS. Soc/Anth. PEAC. Math. Bio. Mass interest meetings: create clear ways for supporters to publicly affiliate w/ campaign. Public media campaign: posters & infographics across campus showcasing all endorsements of movement (pastor in PA?). Develop alumni core: alumni withholding donations. Alumni working on 1) fund 2) letters. NationBuilder is huge for alumni. Have sign-up for network of alums > alumni petition in support of divestment. Get up on NationBuilder before PCM. 100 faculty pledged by mid-March --> 60 faculty by end of November.		
Targets (groups being moved)				All campus allies & neutrals. libera's. ICC/BC. WRC. StuCo greens. Natural sciences & Econ. High-profile alums & donors.		
Actions & events	Pre-PCM: week after retreat: open meeting to turn a lot of people out. intro campaign. etc. 3 orange square ask around come to PCM & come to this meeting next week to prepare for PCM Post-PCM: mass training. Media & comms plan for alumni & faculty to blast out endorsements	Marcie Smith lecture Oct 21 or 22		1st week: low barrier. highly visual. BoM 1: Mark Wallace & student give positive vision & narrative. YouTubed BoM 2: Reiterate positive vision. make direct ask to CN. Public debate between MJ fossil fuel industry supporter. and Swat Admin. Activities Fair.		
Metrics				Train & retain 150 people over course of semester. 75 do more leadership development by end of the fall. 60 faculty pledge. 1.000 alumni e-mails. 10-12 people in the core.		

NAS

Swarthmore Mountain Justice is not a large club. There are five "core organizers" and about 25 participants who regularly attend weekly meetings. But what Mountain Justice lacks in numbers, it makes up in force and strategy. The core organizers work with McChesney, their 350.org mentor, to develop "actions." They have developed a team of three faculty members who counsel them and publicly support them: Mark Wallace, Religion; Betsy Bolton, English; and Lee Smithey, Peace and Conflict Studies. They have scripted talking points and train their members to preach the divestment gospel across campus; one Mountain Justice student told us that the core has worked out conversation scripts to cover both "social justice" and "economics" themed conversations, with questions and arrows indicating follow-on points based on how their interlocutor answers.

They have student committees tasked, for instance, with building coalitions with other student organizations. Another committee "organizes" the faculty to convince them to sign on to divestment. In April 2013, the history department chair Bob Weinberg published a letter in the *Swarthmore Daily Gazette* endorsing Mountain Justice's campaign as a representation of "the best of the liberal arts tradition" and testifying that "There is no greater testament to the value of a liberal arts education than Mountain Justice's campaign for divestment."[497] That same day, 33 history majors released a letter attributing their support for divestment to the lessons they had learned in class about historical oppression.[498] In May, nine religion majors and minors released a similar letter,[499] followed by a letter signed by five religion professors.[500] The English department has done likewise.[501] Thirty professors have signed an open letter, posted to the Mountain Justice website, endorsing divestment as a way to counteract the "unthinkable" damage of "unchecked fossil fuel extraction."[502]

The core group of student activists meets for three hours each week to plan and decide key targets. When Mountain Justice first began, all decisions came to a democratic vote and required unanimity. Occupy Wall Street-style, members would use their fingers to signal on a scale of 0 to 5 how comfortable they were

497   Bob Weinberg, "Op-Ed: History Faculty Supports Divestment in Open Letter," *Swarthmore Daily Gazette*, April 25, 2013. http://daily.swarthmore.edu/2013/04/25/op-ed-history-faculty-supports-divestment-in-open-letter/

498   "Op-Ed: History Students for Divestment," *Swarthmore Daily Gazette*, April 25, 2013. http://daily.swarthmore.edu/2013/04/25/op-ed-history-students-for-divestment/

499   "Religion Department Students for Divestment," *Swarthmore Daily Gazette*, May 7, 2013. http://daily.swarthmore.edu/2013/05/07/religion-department-students-for-divestment/

500   "Op-Ed: Religion Faculty Call for Divestment," *Swarthmore Daily Gazette*, May 19, 2013. http://daily.swarthmore.edu/2013/05/19/op-ed-religion-faculty-call-for-divestment/

501   "Time to Divest: An Open Letter," *Swarthmore Phoenix*, October 23, 2014. http://swarthmorephoenix.com/2014/10/23/12655/

502   "Faculty Open Letter," *Swarthmore Mountain Justice*. http://swatmj.org/faculty/

with a decision; once everyone held up fours and fives, the group moved forward.[503] But unanimity soon became too difficult. Now the five core members call the shots, and the others follow their lead.

The group has high morale and big aims, as evidenced by their 2014-2015 calendar distributed as an example to attendees at a youth climate event. With the help of 350.org, Swarthmore Mountain Justice hopes by the end of the spring 2015 semester to train and retain 150 new student activists, earn the tacit support of another 700-800 students, recruit a majority of the Swarthmore faculty to sign a letter in favor of divestment (with 30-40 lending active campaign support), collect $100,000 from alumni who refuse to contribute to Swarthmore until the college divests, and gather 2,000 emails from alumni lending their support to a divestment petition. They've set out a list of potential allies they hope to court: "liberals, ICC/BC [Intercultural Center/Black Center], WRC [Women's Resource Center], StuCo [Student Council], greens," along with members of the Natural Sciences and Economics departments.

They have lined up Swarthmore religion professor Mark Wallace to speak, and aim to cozy up to Swarthmore alum and board member David Gelber (who produces the TV climate change documentary series *Years of Living Dangerously*) in order to create "space for him to be a climate movement hero." Additional named targets include the Dean of Students Liz Braun; the VP for College and Community Relations and Executive Assistant to the President, Maurice Eldridge; and chairman of the board, Chris Niemczewski. By January 2015, they hoped to ask at least three board members to announce their support, or at least indifference, to divestment.

Mountain Justice scheduled in advance to sit in on the board's February 20-21 meeting with at least 200 students, with preparations for 20-30 social media activists to plaster the Internet with photos and updates, and volunteers to send 100 emails each day to chairman Niemczewski. They also planned to recruit parents to organize and petition the college to divest. They also aim for high-profile coverage from the *New York Times* and *Al Jazeera*. The end goal is to "Escalate until victory. Polarize campus & isolate BoM [Board of Managers]."

---

503 *Institutional Memory Document 2011-2012*, Swarthmore Mountain Justice, August 2012, pg. 23. http://swatmountainjustice. files.wordpress.com/2012/12/mj_institutional_memory_2011-2012_final_draft.pdf

Do the Math

Swarthmore's descent into mobs and scare tactics is perhaps the most extreme example of aggressive sustainability in action, but it embodies the ethos that sustainability embraces. The underlying idea is that fossil fuels desecrate the environment, that this desecration is sinful, and that those who would persevere in their reprehensible ways deserve condemnation. The campaign to divest from fossil fuels manifests the fervor of a religious quest for purity and absolution. It declares the public guilty of oil on their hands.

That campaign started at Swarthmore. When Mountain Justice first launched at Swarthmore in October 2010, Aronoff and her peers were blazing a trail into uncharted territory for the sustainability movement. The sustainability ideology had already succeeded in maneuvering its way into blueprints for campus construction projects and had probed deep into the abysses of employee handbooks and office purchasing guidelines. Up to that point, sustainability had hacked through the overgrowth surrounding the course catalog and carved out a niche in the college curriculum, staked out a homestead in residence life rules, bagged space in trash codes and recycling regulations, and eaten its way into dining hall menus and dish and tray policies.

College and university endowments, though, were a new frontier. Secluded behind the purview of boards of managers and investment committees, and weighed down with the responsibility to churn out money for new buildings, scholarships, financial aid, faculty development, sports complexes, and all the accoutrements required for keeping up with the Joneses, institutional investments proved hard peaks to scale. It's one thing for a college to purchase a couple of battery-operated cars and buy renewable energy credits. It's another to stake the institution's financial future on the forecast that oil will tank long-term. And up to this point, sustainability had itself relied heavily on the institution's finances for purchasing green infrastructure, hiring sustainability directors, retrofitting dorms and classrooms with high-tech heating and power equipment, and other pricy carbon reduction efforts.

Like missionaries pioneering deep into the heart of new lands, planting seeds and hoping for conversions, Mountain Justice began building a network of interested parties who could support and help spread the movement. The students started by partnering with the Responsible Endowments Coalition, a nonprofit that encourages investment into companies deemed environmentally and socially responsible, to learn more about how endowments and investments worked.

Soon Bill McKibben, the environmentalist-journalist who started 350.org, got involved. 350.org, named for the parts-per-million of carbon dioxide in the atmosphere that McKibben deems acceptable, had been primarily focused on organizing public opposition to the Keystone XL pipeline. But McKibben had an epiphany in spring 2012, when he realized he needed to target the fossil   fuel industry itself.

NAS

In his book, *Oil and Honey*, McKibben describes how while reading a *Nature* article about the severe reduction in biodiversity predicted to occur within a few generations, he realized "there was no way, fighting one lightbulb or pipeline at a time, that we could make a dent in that momentum."[504] His strategy to date had been "pretty much the same as everyone else's— go through the political system. Press the president, lobby the Congress, assemble at UN meetings." But, McKibben writes, "I had an idea—that we needed instead to go straight at the fossil fuel industry."[505] According to McKibben, the idea had been growing in his mind for months, apparently independently from Aronoff and the Swarthmore activists, that fossil fuel

Bill McKibben

companies were singlehandedly to blame for the coming global warming catastrophe. Burning oil released carbon dioxide, carbon dioxide was driving global warming, and the companies' profit strategies relied on finding, drilling, selling, and thus enabling people to burn more oil. "It was clear," McKibben writes,

> *that they were planning to wreck the earth. Despite running hundreds of millions of dollars' worth of ads proclaiming their commitment to some kind of sustainable future, their business plans proclaimed in black and white that they were going to take us over the edge.*[506]

That fall, he began bio-diesel bussing between 21 cities on a national speaking tour, "Do the Math." The math, according to McKibben, is simple. Humans can emit no more than 565 gigatons of carbon dioxide in order to stay below the 2 degree Celsius warming that the IPCC declared the maximum safe temperature—"anything more than that risks catastrophe for life on earth," according to McKibben's Do the Math website.[507] "The only problem?" he asks. "Burning the fossil fuel that corporations now have in their reserves would result in emitting 2,795 gigatons of carbon dioxide – five times the safe amount." The bold print in the original emphasizes the danger McKibben detects. As his pictorial equations show, CO2 + $ = a flaming earth. McKibben offers a second mathematical analogy to express the trade-off: "WE > FOSSIL FUELS."[508]

With strobe lights, a rotating line-up of musicians and minor celebrity speakers, upwards of a thousand

---

504 McKibben, Bill, *Oil and Honey*, pg. 140.

505 *Ibid.*

506 McKibben, pg. 149.

507 Do the Math. http://math.350.org/

508 "About the Tour," video, Do the Math. http://math.350.org

NAS

attendees packed into auditoriums, and charismatic, fist-pumping McKibben in center stage warning about "our last best chance to do something," the "talks" took on the verve of a rock concert. Appropriately, in *Rolling Stone*'s August 2nd, 2012 issue, McKibben launched his Do the Math tour with a long article, "Global Warming's Terrifying New Math." "If the pictures of those towering wildfires in Colorado haven't convinced you, or the size of your AC bill this summer, here are some hard numbers about climate change," McKibben opened his piece:

> June broke or tied 3,215 high-temperature records across the United States. That followed the warmest May on record for the Northern Hemisphere – the 327th consecutive month in which the temperature of the entire globe exceeded the 20th-century average, the odds of which occurring by simple chance were 3.7×10^{99}, a number considerably larger than the number of stars in the universe.[509]

He went on to explain his three key numbers—2 degrees, 565 gigatons, and 2,795 gigatons—drawn from a group of financial analysts in the UK. He explained that small personal changes in lifestyle wouldn't do enough fast enough, and that the political process was mired in delays and unreliable outcomes. What would it require to energize a powerful social movement that, of its own volition, adopted substantial changes to shift society towards greener choices and pressure politicians to enact stricter rules? "A rapid, transformative change would require building a movement, and movements require enemies," McKibben mused. "And enemies are what climate change has lacked."[510]

The unconscionable enemy to be scourged, then, is the fossil fuel industry as a whole. "Given this hard math, we need to view the fossil-fuel industry in a new light," McKibben concluded his *Rolling Stone* piece. "It has become a rogue industry, reckless like no other force on Earth. It is Public Enemy Number One to the survival of our planetary civilization."[511]

Igniting the Campus

McKibben's article put fossil fuels on college students' most-wanted lists. The movement caught fire. In November 2012, before McKibben's three-week Do the Math Tour had finished, Unity College in Maine had announced its divestment from fossil fuels.[512] Two months after the tour concluded, in February 2013,

---

509 Bill McKibben, "Global Warming's Terrifying New Math," *Rolling Stone*, July 19, 2012. http://www.rollingstone.com/politics/news/global-warmings-terrifying-new-math-20120719

510 *Ibid.*

511 *Ibid.*

512 Stephen Mulkey, "Unity College Board of Trustees Votes to Divest from Fossil Fuels," *Unity College Sustainability Monitor*, November 5, 2012. http://sustainabilitymonitor.wordpress.com/2012/11/05/unity-college-board-of-trustees-votes-to-divest-from-fossil-fuels/

College of the Atlantic (also in Maine)[513] and Sterling College in Vermont[514] announced their plans to divest. In May, Vermont's Green Mountain College[515] and the San Francisco State University Foundation[516] also announced similar plans. In the two years since McKibben's tour launched, Go Fossil Free has counted thirteen American colleges or universities that have withdrawn their endowments from stocks in fossil fuels.[517] Meanwhile, campaigns popped up around the country, sparked by McKibben, mentored by 350.org staff, and supported by the national Divestment Student Network.

In September 2012, Divest Harvard incorporated as a chapter of the Massachusetts "climate justice" organization Students for a Just and Stable Future (SJSF). It started with six veteran "SJSF-ers" and a handful of new recruits after McKibben made a brief visit to campus. 350.org helped to fund the chapter. Within a few weeks, the tiny team collected more than 1,000 signatures from Harvard students on a petition to divest from fossil fuels. They convinced the Harvard Undergraduate Council to offer a referendum on divestment on the November 2012 student ballot—the first referendum of any kind in six years. It passed by 72 percent.[518]

McKibben's numbers are not the final pronouncement on global warming's effects. Far from worried about runaway warming, many climatologists are actually befuddled by a 15-year "pause" in temperature increases since 1999, despite increasing levels of CO2 in the atmosphere. The following graph from a study in Nature Climate Change shows in the grey shaded area the predictions from the IPCC's

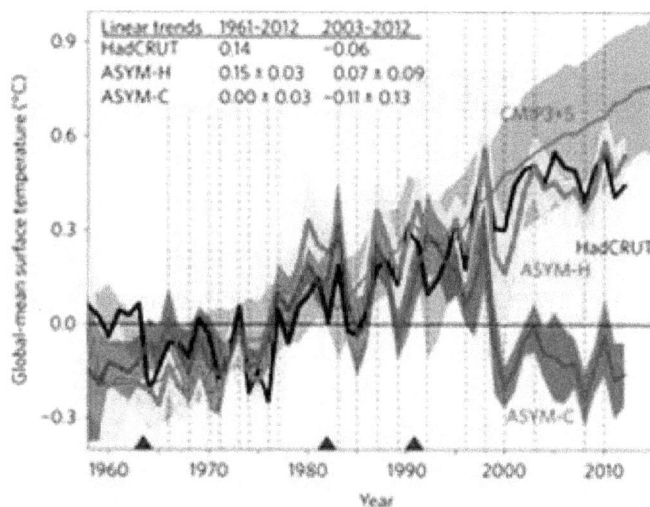

---

513   AP, "College of the Atlantic Sells its Fossil Fuel Investments," *WCSH-6*, March 13, 2013. http://www.wcsh6.com/story/local/2013/03/13/2094768/

514   Terri Hallenbeck, "Sterling College to Divest from Fossil Fuels," *Burlington Free Press*, February 5, 2013. http://blogs.burlingtonfreepress.com/politics/2013/02/05/sterling-college-to-divest-from-fossil-fuels/

515   "Green Mountain College Divests from Fossil Fuels," *Burlington Free Press*, May 14, 2013. http://www.burlingtonfreepress.com/article/20130514/GREEN/305140020/Green-Mountain-College-divests-from-fossil-fuels

516   Jamie Henn, "San Francisco State University Divests from Coal and Tar Sands!" Go Fossil Free, June 11, 2013. http://gofossilfree.org/san-francisco-state-university-divests-from-coal-and-tar-sands/

517   "Divestment Commitments," Go Fossil Free. http://gofossilfree.org/commitments/

518   "Campaign Narrative," Divest Harvard. http://divestharvard.com/resources/campaign-narrative/

NAS

Coupled Model Intercomparison Project Phase 5 (used in the Fifth Assessment Report). The black line underneath shows the real surface temperatures.[519]

Nor is global warming universally perceived as so harmful as McKibben fears. Robert Mendelsohn, the Edwin Weyerhaeuser Davis Professor of Forest Policy at Yale, and himself an adherent to the theories of anthropogenic global warming, calculates that up to 4 degrees Celsius of warming would actually benefit, not harm, most of earth's ecosystems, and that increased levels of CO2 have led to increased greenery.[520] And McKibben's own presentation of the data shows a more promising picture than he lets on. Even if 2 degrees were the maximum safe level of warming, and even if the cumulative gigatons of unburned fossil fuels could set us over the 2-degree mark, McKibben himself calculates that the 6 largest companies currently hold enough oil to cause, at maximum, a 0.5 degree temperature rise, or only 25 percent of the "carbon budget."

In response to the growing divestment movement, the *New York Times* offered a sympathetic front-page article, "To Stop Climate Change, Students Aim at College Portfolios."[521]

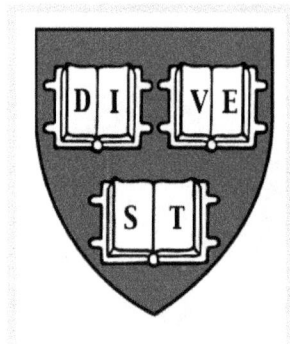

Divest Harvard offers a spin on Harvard's crest.

Divest Harvard ran a textbook strategy: they began meeting with the administration, recruited faculty to speak for them, got a meeting with members of the Harvard Corporation Committee on Social Responsibility, raised money and hired student staff, started a newsletter, and held rallies. They reached out to Harvard's graduate schools, and in spring 2013, Harvard Law School students voted by 67 percent to support divestment. [522]

During the 2013-2014 school year, would-be divesters grew bolder. They wrote op-eds for the *Harvard Crimson*, repeating McKibben's talking points about moral urgency and the blameworthiness of fossil fuels. "The goals of the movement are twofold," undergraduates Joseph G. Lanzillo, Chloe S. Maxmin, and Pennilynn R. Stahl wrote:

> First, we are highlighting the destructive and deeply irresponsible practices of the fossil fuel industry.
> If an industry refuses to support the future of life on Earth, then we must refuse to support the

---

519   Masahiro Watanebe, et. al., "Contribution of Natural Decadal Variability to Global Warming Acceleration and Hiatus," *Nature Climate Change*, August 31, 2014.   http://www.nature.com/nclimate/journal/vaop/ncurrent/full/nclimate2355.html

520   Amy Athey McDonald, "Environmental Dollars and Sense: Q&A with Robert O. Mendelsohn," *Yale News*, April 21, 2014. http://news.yale.edu/2014/04/21/environmental-dollars-and-sense-qa-robert-o-mendelsohn

521   Justin Gillis, "To Stop Climate Change, Students Aim at College Portfolios," *New York Times*, December 4, 2012. http://www.nytimes.com/2012/12/05/business/energy-environment/to-fight-climate-change-college-students-take-aim-at-the-endowment-portfolio.html?pagewanted=all&_r=1&

522   "Campaign Narrative," Divest Harvard. http://divestharvard.com/resources/campaign-narrative/

NAS

industry. Second, divestment sends a strong political message that climate change is a crisis that needs to be addressedtoday; it creates the public platform necessary to hold politicians accountable to the voices of their constituents.[523]

In September, 100 students and alumni demonstrated outside Massachusetts Hall and demanded a meeting with President Drew Faust by October 1st.[524] In response, Faust averred that the primary purpose of university money was to advance academics, not political causes, and that divestment would harm the endowment's performance. Her statement offered a resounding defense of the university's academic essence:

> *Harvard is an academic institution. It exists to serve an academic mission — to carry out the best possible programs of education and research. We hold our endowment funds in trust to advance that mission, which is the University's distinctive way of serving society. The funds in the endowment have been given to us by generous benefactors over many years to advance academic aims, not to serve other purposes, however worthy. As such, we maintain a strong presumption against divesting investment assets for reasons unrelated to the endowment's financial strength and its ability to advance our academic goals.*

"We should be very wary of steps intended to instrumentalize our endowment in ways that would appear to position the University as a political actor rather than an academic institution."

- Harvard President Drew Faust

> *We should, moreover, be very wary of steps intended to instrumentalize our endowment in ways that would appear to position the University as a political actor rather than an academic institution. Conceiving of the endowment not as an economic resource, but as a tool to inject the University into the political process or as a lever to exert economic pressure for social purposes, can entail serious risks to the independence of the academic enterprise. The endowment is a resource, not an instrument to impel social or political change.*[525]

---

523   Joseph G. Lanzillo, Chloe S. Maxmin, and Pennilynn R. Stahl, "'Extraordinarily Rare' Is not an Excuse," *Harvard Crimson*, September 24, 2013. http://www.thecrimson.com/article/2013/9/24/rare-is-not-enough/

524   "Thank You for an Awesome Demonstration!" Harvard Alumni for Divestment. http://us6.campaign-archive2.com/?u=fa136 1b4ea&id=f283e87d91&e=UNIQID

525   Drew Faust, "Fossil Fuel Divestment Statement," Office of the President, Harvard University, October 3, 2013. http://www. harvard.edu/president/fossil-fuels

NAS

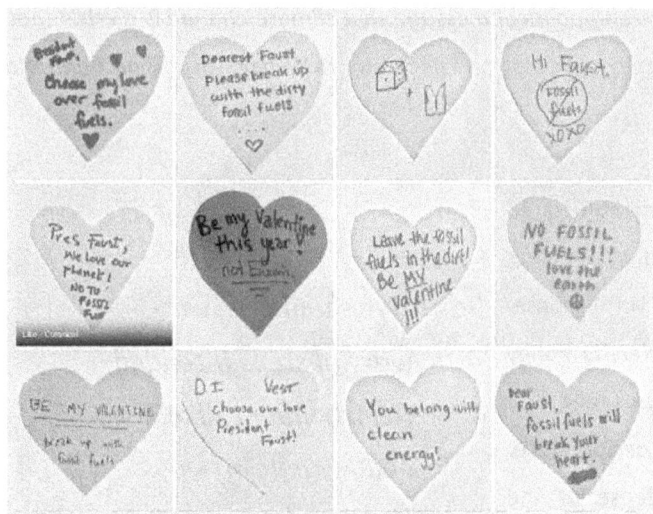

Harvard students asked President Drew Faust to "break up with fossil fuels" on Valentine's Day.

In response, Divest Harvard held a rally with four Harvard professors and the undergraduate student body president as speakers.[526] On Valentine's Day they mailed 100 notes to Faust, asking her to "break up" with fossil fuels. Faust received construction paper hearts marked, "Be my Valentine this year, not Exxon's," and "You belong with clean energy!"

Then on April 30, 2014, 30 students surrounded Faust's office, linking arms across the entrance, blockading the doors, and vowing to stay until Faust agreed to an open meeting on divestment. When one of the students, Brett Roche ('15) was arrested by the police for his action, he was hailed by Divest Harvard as a hero. All this fueled support for divestment. As of January 2015, 1,044 alumni have signed a petition asking their alma mater to divest.[527] Two hundred twenty-nine Harvard professors have signed a similar pledge.[528]

The Orange Square Movement

In the two years since McKibben branded the fossil fuel industry as the number one enemy to Earth and its inhabitants, more than 400 fossil fuel divestment campaigns have broken out at colleges and universities worldwide. McKibben's network, Go Fossil Free, developed a divestment textbook,[529] investment guides,[530] recruitment toolkit,[531] sample divestment resolutions,[532] and fact sheets[533] to arm students for their campus battles. A sample poster shows a pumpkin-orange sheet sliced on the right by a single white smokestack tattooed with the names of all the evils it emits: climate change, mass extinction,

---

526 "Divest Harvard Presents: Forum on Fossil Fuel Divestment," Divest Harvard. http://divestharvard.com/divest-harvard-presents-forum-on-fossil-fuel-divestment/

527 "Alumni," Divest Harvard. http://divestharvard.com/alumni

528 "Harvard Faculty: Join Us," Harvard Faculty for Divestment. http://www.harvardfacultydivest.com/add-your-name/

529 "Fossil Free: A Campus Guide to Fossil Fuel Divestment," Go Fossil Free. https://s3.amazonaws.com/s3.350.org/images/350_FossilFreeBooklet_LO4.pdf

530 "A Complete Guide to Reinvestment," Go Fossil Free. https://s3.amazonaws.com/s3.350.org/images/Reivestment_Guide.pdf

531 "Recruitment Toolkit," Go Fossil Free. https://s3.amazonaws.com/s3.350.org/images/FossilFreeRecruitmentToolkit.pdf

532 "Fossil Fuel Divestment Resolutions," Go Fossil Free Google Drive. https://drive.google.com/folderview?usp=sharing&pli=1&id=0B0dtRhE8YprQZXNFZWNobUl3V0E&ddrp=1#

533 "Climate Impacts Fact Sheets," Connect the Dots. http://www.climatedots.org/factsheets/

dirty energy, cognitive dissonance, corrupt government, 333rd month of above average temperatures, sea level rise, Sandy (that is, Hurricane Sandy that hit the Northeast in October 2012), children with asthma, the Niger delta (the Nigerian coast facing floods and erosions; lowercase in original), Colorado fires, cancer alley, and, finally, at the base of all these, in thick orange capitals, FOSSIL FUELS.[534] How "cognitive dissonance" results from burning oil is perhaps a mystery meant to be pondered by the viewer.

The color orange in the poster is key in light of the broader divestment movement. The top half of Go Fossil Free's homepage, bright orange above the grey texts beneath, sets forth in grey and white block letters the main values of the movement. The screen reads, "Divesting from Fossil Fuels is," and a scrolling list enumerates all the positives of what divestment aims to be: "Smart. Ethical. Forward-Thinking. Sustainable. Prudent. the right thing to do."[535] Vaguely shadowed behind the orange screen is a heartrending image of a child climbing over the rubble remains of a wood and tarp hut, presumably his home destroyed by a global-warming induced storm.

Orange is the standard color for fossil fuel divestment campaigns.

The color is repeated again in Go Fossil Free's divestment textbook, "Fossil Free: A Campus Guide to Fossil Free Divestment," where the cover features black rubber-stamped letters atop a painted orange X, the motif that serves as Go Fossil Free's logo.[536] Swarthmore Mountain Justice released its own guide, "Fossil Fuel Divestment 101," complete with a list of "the sordid sixteen" worst companies, where orange is again

534 Lisa Purdy, "Divest From...Poster," Go Fossil Free. https://s3.amazonaws.com/s3.350.org/images/Divestfrom_Poster_LisaPurdy.pdf

535 Go Fossil Free. http://gofossilfree.org/

536 "Fossil Free: A Campus Guide to Fossil Fuel Divestment," 350.org. https://s3.amazonaws.com/s3.350.org/images/350_FossilFreeBooklet_LO4.pdf

the dominant color.[537]

Even We Are Powershift, a "grassroots-driven online community" for college students who hope to shift American power sources away from fossil fuels and towards renewable energy sources, throws a dash of orange into its predominantly teal color scheme. On the homepage, on the scrolling slides advertising the Powershift Convergence from October 2013, when 10,000 students coalesced in Pittsburgh to hear 200 speakers and musicians rally around divestment, a second slide shows the transformation that Powershift and its allies aim for. There, in another pictorial equation, blue chimneys belching puffy black smoke behind a black cupola-topped college hall meet a

Even PowerShift added orange to its homepage.

zealous student waving a sign before his now-orange college hall, and the result after the "equal" sign is a freshly painted green hall topped with solar panels. The changing colors of the campus indicate that the orange symbolizes something that happens inside the campus, not outside in the economy. Perhaps it represents the institutional version of a change of heart as colleges repent and turn away from the fossil fuel holdings in their endowments.

Three Swarthmore students, Sachie Hopkins Hayakawa, Sally Brunner, and Lauren Ressler, explain the significance of the color in a February 2013 article on Powershift's website, written just before Mountain Justice organized a march on Swarthmore's administrative building:

> This afternoon, as students take action at Swarthmore College we will be wearing orange squares pinned to our chests. We have chosen to wear this symbol today in solidarity with other student power movements internationally — most notably the Quebec Student Movement.[538]

Out of protest for rising student debt, students in the Quebec Student Movement had taken to wearing red squares in honor of the phrase, "carrément dans le rouge," or "squarely in the red." In 2012, as Quebec students donned their red squares and marched the streets, businesses hung red squares in their windows in shows of solidarity and fabric shops sold out of their red stock. That was the kind of social movement with complete cultural saturation that Mountain Justice was hoping to emulate. Hayakawa, Brunner, and Ressler went on to explain that the color orange not only distinguished them from the

---

537   Swarthmore Mountain Justice, *Fossil Fuel Divestment 101*, May 2013. http://swatmountainjustice.files.wordpress.com/2013/04/fossil-fuel-divestment-101_may-2013.pdf

538   Sachie Hopkins Hayakawa, Sally Brunner, and Lauren Ressler, "A New Symbol? The Orange Square," *Powershift*, February 25, 2013. http://www.wearepowershift.org/blogs/new-symbol-orange-square

Quebec anti-debt movement, but it also distanced them from the traditional environmental movement. Their new movement had a broader focus that bled into spheres of sociology, economics, and more:

> *We have chosen the color orange, rather than green, to reframe our movement's scope as much larger than an environmental issue. This is not a single-issue movement. This is a space where environmental justice, climate justice, and economic justice have come into contact. We understand that we will not win the fight against the fossil fuel industry without confronting racism, classism, homophobia, and other systems of oppression in our movement spaces. At this convergence we have begun conversations about intersectionality and historical responsibility on an international scale.[539]*

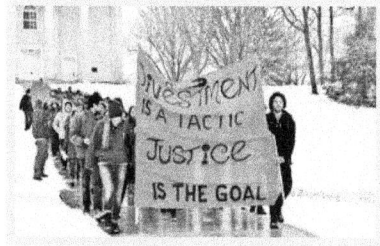
Middlebury College students march on campus.

The rhetoric sounds like something their Swarthmore co-agitator Watufani Poe might have shouted from the mic at the board of managers meeting. These three students testify to divestment's broader social goals—confronting racism, classism, homophobia, and "systems of oppression"—and to the ease with which the sustainability movement glides from clean energy advocacy to a hard left social agenda. What racism has to do with extracting oil is an enthymematic riddle, with the apparent premise that environmental oppression somehow mimics or reinforces racial oppression. As Barry Commoner put forth, "It's all interconnected."

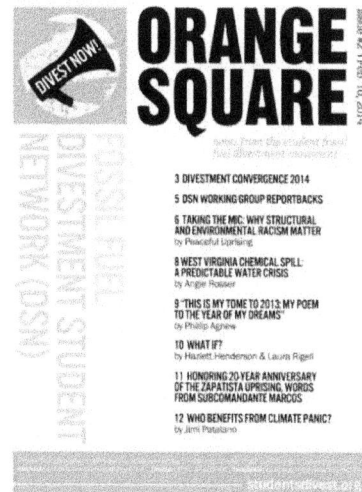

The orange square has become a rallying symbol in the divestment movement. *Orange Square* is the name of an online publication of the Fossil Fuel Divestment Student Network. Students affiliated with "Beyond Coal" at the University of Illinois-Urbana Champaign held an "Orange Square Day" on September 19, 2013, chalking orange squares on sidewalks and handing out squares of orange felt on safety pins.[540] In March 2013, under heavy snowfall, 125 hearty Middlebury College students marched across campus wearing orange square badges and carrying an oversized orange square banner that read, "Divestment is a tactic. Justice is the goal." As they marched they chanted, "Money for students' education, not for climate devastation. Money for homes and education, not for war and exploitation."[541] The irony, of course, is that were Middlebury to divest, it would lose money and thus

---

539 *Ibid.*

540 "Beyond Coal Orange Square Day," UIUC Beyond Coal, September 25, 2013. http://www.uiucbeyondcoal.com/2013/09/beyond-coal-orange-square-day.html

541 Bronwyn Oatley, "Students March for Divestment," *The Middlebury Campus*, March 6, 2013. http://middleburycampus.com/article/students-march-for-divestment/

have substantially *less* money for homes and education.

Talking Money

McKibben's movement, Go Fossil Free, asks institutions to vow to refrain from investing in any of the 200 companies on the Fossil Free Index's list of worst fossil fuel companies,[542] and to sell off within five years any stock the universities hold in these companies.[543] The issue gets tricky, though, when institutions invest in commingled funds, and they don't have direct control over which companies the fund invests in. Does a full fossil fuel purge require selling off those as well? That, combined with the fact that fossil fuels tend to provide good market returns over the long term, can make divestment pricey. Just before Mountain Justice interrupted him from speaking, the Swarthmore chairman of the finance committee, Chris Niemczewski, was about to explain that divestment would cost the college $200 million over the course of 10 years.[544]

Divestment activists counter that divestment is actually a smart financial strategy, because once global warming becomes internationally realized and nations begin enacting stringent carbon caps, fossil fuel companies will be forced to quit extracting oil and to leave most of their bounty in the ground. This means that current fossil fuel stocks enjoy an artificially high value that will plummet once these carbon caps come into effect. This "carbon bubble" will eventually collapse, and the institutions that divest will be best prepared to ride out the wave of economic upheaval that follows. Would-be divesters frequently quote a report from the Aperio Group, which acknowledged that eliminating categories of stocks inevitably increases the risk of a portfolio, but argues that the additional risk of getting rid of fossil fuels is a mere 0.0002 percent.[545]

But Aperio Group calculates only a portion of the risk associated with divestment. It compares a portfolio that holds stock in every company in the Russell 3000 Index with a custom carbon-free portfolio that holds stock in every Russell 3000 company except for 13 fossil fuel companies. In fact, few college endowments simply invest in these index funds, because they earn a merely average market rate of return. Savvy investors aim to pinpoint the most valuable stocks and outperform the market rate. Swarthmore's finance chair, Chris Niemczewski, tried to explain this to the students before he was interrupted by Mountain Justice. Later he explained to the *Swarthmore Daily Gazette* that Swarthmore hires investment managers to actively manage much of the endowment, and that these funds have earned between 1.7

---

542   "The World's Top 200 Public Companies," Fossil Free Indexes. http://fossilfreeindexes.com/the-carbon-underground-2014/

543   "Fossil Fuel Divestment: Colleges and Universities," Go Fossil Free. http://campaigns.gofossilfree.org/efforts/fossil-fuel-divestment-colleges-universities/petitions/new

544   Andrew Karas, "Swarthmore Pegs Cost of Divestment at $200 Million Over 10 Years," *Swarthmore Daily Gazette*, May 9, 2013. http://daily.swarthmore.edu/2013/05/09/college-pegs-cost-of-divestment-at-200-million-over-10-years/

545   Patrick Geddes, "Do the Investment Math: Building a Carbon-Free Portfolio," Aperio Group, 2013. http://www.aperiogroup.com/system/files/documents/building_a_carbon_free_portfolio.pdf

percent and 1.8 percent above the average market rate of return. Because these funds hold investments from Swarthmore as well as from other institutions, Swarthmore cannot simply order the funds managers to divest from fossil fuels and invest elsewhere. Divestment would require creating custom funds for Swarthmore alone (which carry higher fees for managers). And because, to the board's knowledge, no actively managed carbon-free funds exist, Swarthmore could no longer actively manage its investments and would be forced to invest simply in index funds. Swarthmore would give up its 1.7-1.8 percent advantage in its actively managed funds, plus pay fees for hiring managers to create these custom funds. These are the losses that Niemczewski estimated would reach $200 million within 10 years.[546]

Another consideration is the booming business of fossil fuels. Divestment activists may hope for a "carbon bubble" in the future, but at present, despite a recent bump caused by rapid production of oil, fossil fuels are valuable. Hydraulic fracturing has opened new sources of energy that have discredited fears of "peak oil"—the hypothetical moment when extraction of oil goes into irreversible decline. In fact, as energy prices rose, the oil industry devised newer and better ways to tap reserves that were previously unreachable. As a result of these technological innovations and changes in the market, the United States is now the largest energy producer in the world.[547]

The costs of divestment are a substantial obstacle. *Inside Higher Ed*'s annual survey of chief financial officers at colleges and universities shows that in 2014, only 6 percent were interested in divestment.[548] The previous year, 75 percent of CFOs disagreed or strongly disagreed with the student push to divest endowments from fossil fuels, and 56 percent agreed that colleges should "focus on financial issues (not ethical or political ones)" in their decisions to invest the endowment.[549]

The financial costs are a large reason why the colleges who have announced their divestments are mostly small, impecunious colleges with small endowments and, consequently, smaller losses. Five of the twenty institutions have endowments of less than $5 million. Two of these do not reach even $1 million. By comparison, of the 1,141 colleges and universities ranked by *U.S. News and World Report*, the average endowment size is $329.9 million. Harvard has the largest endowment at $30 billion.[550]

---

546   Karas, "Swarthmore Pegs Cost of Divestment at $200 Million Over   10 Years."

547   Russell Gold, "Fracking Gives U.S. Energy Boom Plenty of Room to Run," *Wall Street Journal*, September 14, 2014. http://online.wsj.com/articles/fracking-gives-u-s-energy-boom-plenty-of-room-to-run-1410728682

548   Ry Rivard, "Sustainability, Divestment and Debt: A Survey of Business Officers," *Inside Higher Ed*, July 18, 2014. https://www.insidehighered.com/news/survey/sustainability-divestment-and-debt-survey-business-officers

549   Doug Lederman, "CFO Survey Reveals Doubts About Financial Sustainability," *Inside Higher Ed*, July 12, 2013. https://www.insidehighered.com/news/survey/cfo-survey-reveals-doubts-about-financial-sustainability

550   Devon Haynie, "Universities with the Largest Financial Endowments," *U.S. News and World Report*, October 1, 2013. http://www.usnews.com/education/best-colleges/the-short-list-college/articles/2013/10/01/universities-with-the-largest-financial-endowments-colleges-with-the-largest-financial-endowments

NAS

Only three institutions with plans to divest have endowments above that average: the New School at $214 million,[551] the University of Dayton, at $442 million,[552] and Stanford, with an endowment of almost $18.7 billion—though Stanford agreed to divest only from coal, not from oil.[553]

Figure 27. Endowment Sizes of Institutions that Divested from Fossil Fuels

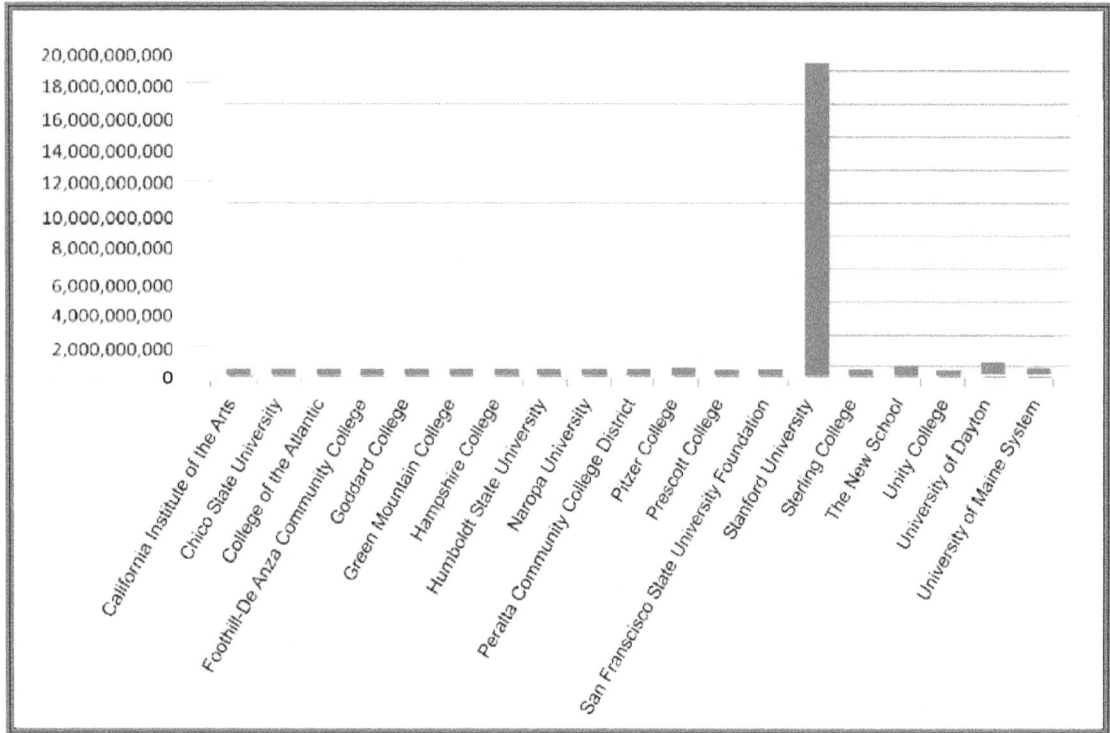

---

551   The New School, Sortable Table: College and University Endowments, 2012-3, *Chronicle of Higher Education*, January 28, 2014. http://chronicle.com/article/Sortable-Table-College-and/144241/

552   University of Dayton, Sortable Table, *Chronicle of Higher Education*.

553   Stanford University, Sortable Table, *Chronicle of Higher Education*.

Figure 28. Endowment Sizes of Institutions (Minus Stanford) that Divested from Fossil Fuels

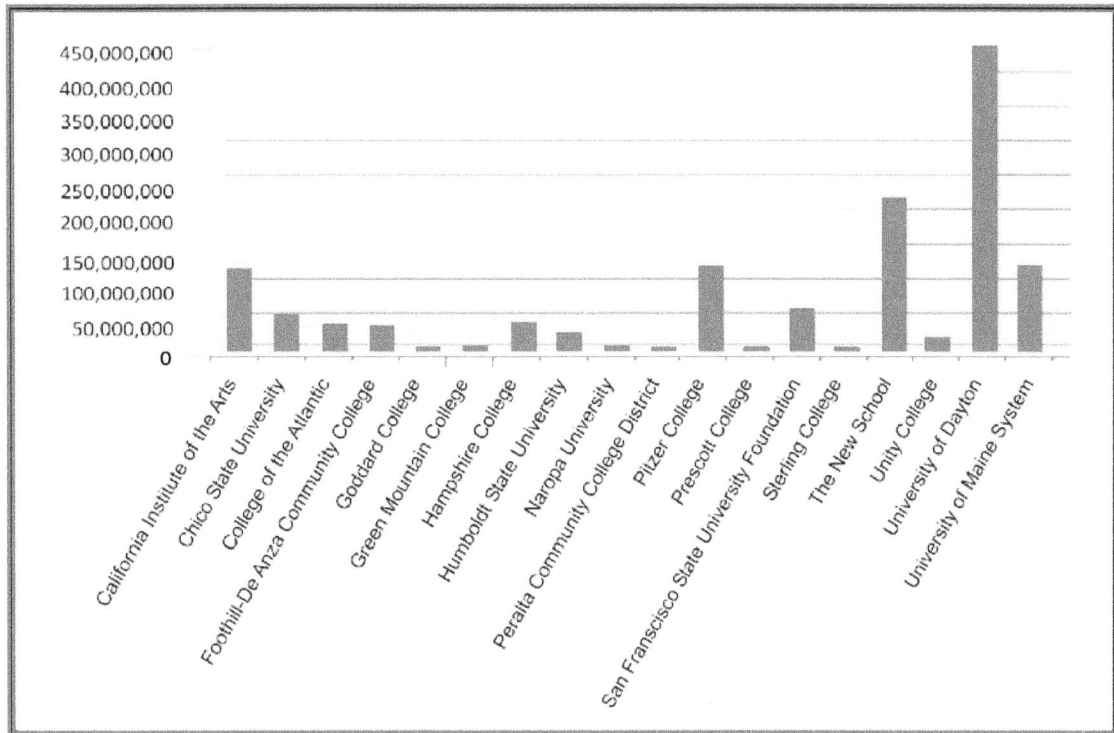

Other colleges that declined to divest came up with creative ways to spare the costs of divestment while attempting to placate student and alumni interested in casting out fossil fuels. Citing the substantial economic costs to divesting Harvard's endowment, President Faust in April 2014 announced Harvard's adoption of the UN's Principles for Responsible Investment to integrate sustainability concerns into future investment decisions, and Harvard's signing onto the Carbon Disclosure Project, which pressures companies to release data on their carbon emission in the hopes that transparency will force them to clean up their operations.[554]

Tufts University declined in March 2014 to divest "at this time"—leaving open the possibility of divestment in the future—after a consultant determined that divestment would affect 60 percent of the endowment and result in losses of $75 million. Instead, Tufts agreed to start a Sustainability Fund that would not invest in any fossil fuels. Donors wishing to keep their money "clean" could donate to this fund.[555]

---

554 Drew Faust, "Confronting Climate Change," Office of the President, Harvard University, April 7, 2014. http://www.harvard.edu/president/news/2014/confronting-climate-change

555 Ben Weilerstein, Devyn Powell, and StudentNation, "Tufts Students Say Fear Held the University Back from Fossil Fuel Divestment," *The Nation*, March 5, 2014. http://www.thenation.com/blog/178709/tufts-students-say-fear-held-university-back-fossil-fuel-divestment#

NAS

Political Bankruptcy

The costs of divestment preclude most campaigns from getting much further than a few meetings with the board, a series of op-eds, and one good Swarthmore-esque rally. Administrators toss a bone to student activists—either out of real sympathy for and agreement with the students' motives, or from a strategic calculation to conserve political capital. But then they deflect the divestment drama by throwing up their hands. They are, after all, the ones accountable for the financial preservation of the institution. At Swarthmore, President Chopp entertained proposals from Swarthmore's pro-divestment Mountain Justice chapter but then refrained from weighing in directly, instead citing endowment guidelines requiring Swarthmore to "manage the endowment to yield the best long-term financial results, rather than to pursue social objectives."[556] At Bowdoin College, Barry Mills met with students pushing a proposal to divest, and when he vetoed their proposal, he let them down gently: "I would never say never."[557]

What is the purpose of divestment, then, besides creating an ineffective student frenzy? Some say the goal is to exert social pressure to force the fossil fuel industry to change its ways or else fizzle. If enough colleges strip their endowments of fossil fuel stocks, then perhaps the targeted 200 companies might suffer some financial loss.

The effectiveness of the tactic, though, is doubtful. Divestment, like a boycott, requires a critical mass in order to generate sufficient pressure. But in boycotts, the drop in demand for the boycotted product is real. The boycotting consumer isn't replaced by another consumer—unless there's a counter boycott. But when one college divests from fossil fuel companies, any number of investors will eagerly buy up the stocks. The company scorned by one college is eagerly welcomed by another.

What is the purpose of divestment, then, besides creating an ineffective student frenzy? Some say the goal is to exert social pressure to force the fossil fuel industry to change its ways or else fizzle. If enough colleges strip their endowments of fossil fuel stocks, then perhaps the targeted 200 companies might suffer some financial loss.

These reasons lead a number of environmentalists to consider divestment campaigns economically foolish. In a 2013 *New York Times* "Room for Debate" series, environmental activists discussed whether divestment movements worked.[558] Three of the five contributors suggested that divestments didn't. As

---

556   Rebecca Chopp, "Chopp: Op-Ed on Divestment," *Swarthmore Daily Gazette*, April 19, 2012. http://daily.swarthmore.edu/2012/04/19/12697/

557   Marisa McGarry, "Mills Says College Will Not Divest from Fossil Fuels," *Bowdoin Orient*, December 7, 2012. http://bowdoinorient.com/article/7814

558   Bill McKibben, "Turning Colleges' Partners into Pariahs," *New York Times*, Room for Debate, January 27, 2013. http://www.nytimes.com/roomfordebate/2013/01/27/is-divestment-an-effective-means-of-protest/turning-colleges-partners-into-pariahs

NAS

two Johns Hopkins students wrote, divesters "are merely absolving themselves of direct moral culpability while hiding their heads in the sand."[559] Instead, they recommended shareholder advocacy from within. Colleges could buy up stocks in fossil fuel companies, and then as major owners of the companies, begin petitioning for greater investment in solar, wind, and bio-fuel energy.

The truth of these objections has forced proponents of divestment to cast about for another justification. Increasingly, they've settled on the principle of public shaming. McKibben put this argument more bluntly than most in his own contribution to the *New York Times* debate: "Turning Colleges' Partners into Pariahs."[560] The "partners" in this case are the corporations with whom colleges have partnered in crime as they "wreck the climate." The way to take down these carbon-emitting corporations is not to ruin their bottom lines, McKibben argues, but to "revoke the social license of these firms." The goal is not just to stop fossil fuels from getting burned, or to drain the finances of the industry, but to discredit and embarrass them, and to prevent the public from trusting what they said.

McKibben sees fossil fuel companies as public enemies not just for their ability to drill and enable the burning of oil and gas, but because they have a lot of money available to spend on political lobbying. "Left to our own devices," McKibben suggested in his *Rolling Stone* piece, "citizens might decide to regulate carbon." He cited a poll that found that "nearly two-thirds of Americans would back an international agreement that cut carbon emissions 90 percent by 2050."[561] But Americans weren't left to their own devices, he held, because the Koch brothers, who "made most of their money in hydrocarbons" and who "know any system to regulate carbon would cut those profits," hold "a combined wealth of $50 billion" and "reportedly" planned to spend $200 million on the 2012 election.[562] And in 2009, the U.S. Chamber of Commerce "surpassed both the Republican and Democratic National Committees on political spending" and entrusted 90 percent of their funds to the campaigns of "GOP candidates, many of whom deny the existence of global warming."[563]

Aronoff, the Swarthmore Mountain Justice student, echoed McKibben in her piece for the *Times*:

> Students, alumni and faculty have unique access to their universities' moral suasion and material wealth: the collective $400 billion held in university endowments. While these funds alone are unlikely to drive down the stock prices of companies like Chevron and ExxonMobil, observers of both

559 David Israel and Nikko Price, "Change from Within Is More Effective," *New York Times*, Room for Debate, January 27, 2013. http://www.nytimes.com/roomfordebate/2013/01/27/is-divestment-an-effective-means-of-protest/change-from-within-is-more-effective

560 McKibben, "Turning Colleges' Partners into Pariahs."

561 McKibben, "Global Warming's Terrifying New Math."

562 *Ibid.*

563 *Ibid.*

NAS

the nation's history and our current political situation know that mass movements and money are what talk in Washington.[564]

By shaming corporations that emit too much carbon, divesters hope to decimate those corporations' credibility and silence their lobbyists, whom McKibben blames for preventing environmental regulations from passing through Congress. Divestment is then the tool of the social architect looking to reshape the norms of social assumptions.

Hence Divest Harvard is shifting its tactics. Eventually, Harvard students would like their institution to

- freeze any new investments in fossil fuel companies

- divest direct holdings (currently $17.3 million) from the top 200 publicly traded fossil fuel companies

- divest indirect holdings in the top 200 fossil fuel companies within 5 years, and reinvest in socially responsible funds.[565]

But they know that's too much for Harvard to swallow, and so, according to Ben Franta, a Ph.D. student in Applied Physics and one of the original members of Divest Harvard, activists have made a calculated decision to focus right now only on divestment from direct holdings (rather than indirect holdings) in fossil fuels. Only 3 percent of Harvard's endowment (about $1 billion) is in direct holdings, and of this, only about 3 percent is in fossil fuels. In sum, Divest Harvard asking Harvard to divest 3 percent of 3 percent, or .0009 of the endowment. How much would a .0009 drop in Harvard's endowment investments harm the fossil fuel companies? Probably not much. But, Franta explained to us, "That's fine." The goal is less about bankrupting the fossil fuel industry financially than it is about bankrupting them politically.

At first, this rhetoric seems to differ greatly from the cries of moral duty to wash our hands of oil and its money. This version of divestment rests on a strictly utilitarian calculus. The divestment movement won't succeed by starving the fossil fuel industry of capital, so it will seek to starve it of political power. The idea is to generate political pressure "because of the huge discussion that it creates and the symbolic action of labeling some action socially irresponsible," as Franta explained.

He believes that oil companies fund "disinformation campaigns" that pacify the public with studies claiming to find no evidence for global warming, an argument that Harvard science historian Naomi Oreskes has advanced in her book *Merchants of Doubt*. Franta also commented that these oil companies

---

564   Kate Aronoff, "A Powerful Way to Galvanize Protest Over Climate Change," *New York Times*, Room for Debate, January 27, 2013. http://www.nytimes.com/roomfordebate/2013/01/27/is-divestment-an-effective-means-of-protest/a-powerful-way-to-galvanize-protest-over-climate-change

565   Divest Harvard. http://divestharvard.com/

bribe politicians with high-dollar campaign contributions—particularly from Charles and David Koch—in order to buy easy environmental regulations. "These disinformation campaigns make students really mad," Franta said. He went on,

> *They (the students) lose out. There is a self-interest element here. The Apartheid divestment campaign appealed to students' sense of justice, that people elsewhere were being mistreated. The fossil fuel divestment campaign also appeals to their sense of justice, but also to their self-interest. They're the ones being hurt.*

The moral duty, then, is not so much to purge ourselves of oil, but to purge the oil companies of social standing. It's not our sin in using gasoline that concerns divesters. It's the industry's sin of mere existence that offends their moral sensibilities.

Environmental Justice

Ben Franta, the Harvard grad student, referenced Apartheid as the touchstone for fossil fuel divesters. McKibben, too, links the Apartheid divestment movement in the 1970s to the modern-day fossil fuel divestment movement as part of a powerful heritage of the people's power to effect moral change. His *Rolling Stone* article drew inspiration from the group-movement tactics of the previous generation's divestment:

> *Once, in recent corporate history, anger forced an industry to make basic changes. That was the campaign in the 1980s demanding divestment from companies doing business in South Africa. It rose first on college campuses and then spread to municipal and state governments; 155 campuses eventually divested, and by the end of the decade, more than 80 cities, 25 states and 19 counties had taken some form of binding economic action against companies connected to the apartheid regime.*[566]

McKibben's argument—that fossil fuel divestment inherits the moral urgency and the pragmatic tactics of Apartheid divestment—got a lift when Desmond Tutu, the Nobel Peace Prize winner and South African anti-Apartheid leader, endorsed fossil fuel divestment and joined him on his Do the Math Tour. The title of Tutu's April 2014 opinion piece in the *Guardian* sums up his thoughts: "We Need an Apartheid-Style Boycott to Save the Planet."[567]

McKibben's protégés get the connection. A Swarthmore Mountain Justice representative writing with two other activists from Haverford College and Bryn Mawr at the *Swarthmore Daily Gazette* compared the horrors of racially segregated South Africa to the oncoming catastrophes of global warming:

---

566   McKibben, "Global Warming's Terrifying New Math."

567   Desmond Tutu, "We need an Apartheid-Style Boycott to Save the Planet," *Guardian*, April 10, 2014. http://www.theguardian.com/commentisfree/2014/apr/10/divest-fossil-fuels-climate-change-keystone-xl

Like divestment from apartheid, the fossil fuel divestment movement is a student-led movement challenging entrenched injustices and demanding real solutions. [568]

Students at Hamilton College, working with Al Gore's business partner and former Hamilton board member David Blood on a divestment campaign there, made much of Hamilton's divestment movement during Apartheid, when their forerunners pressured the administration to divest from companies with ties to South Africa.[569] Hamilton's president at the time had responded by suspending the twelve students who camped out in his office and by destroying the shanties students had built and lived in intending to express solidarity with impoverished South Africans. "The strong media frenzy that accompanied the anti-apartheid movement helped to end the atrocity," two Hamilton students wrote in *The Spectator*, Hamilton's student newspaper. "By placing pressure on such institutions, divestment generates media and financial pressure on the industry or nation it is focused on." Perhaps they thought that if they also got suspended, or arrested like Harvard student Brett Roche, this, too, would generate media attention.

The Apartheid reference demonstrates the flip side to the environmental justice argument. The rub is not merely that students get misled by industry-sponsored "disinformation campaigns," or that the political system is morally hampered by Chevron lobbyists. Global warming is thought to disproportionately affect poor, often African, countries, though it is the richer, typically Western, nations that do most of the carbon emitting. Swarthmore Mountain Justice's *Divestment 101* handbook explains:

> *Climate change and fossil fuel extraction most severely impact populations that are already marginalized. In the United States that means low income communities and communities of color who have been fighting the pollution and political corruption of fossil fuel companies for decades. Globally, climate change disproportionately impacts poor communities in the global south - those least responsible for climate change.* [570]

William Lawrence, a Swarthmore graduate who was among the founding members of Swarthmore Mountain Justice, explained in an interview that the idea of environmental justice "captures the fact that climate change is a social justice issue." He continued,

> *It is an issue which is related to the racism that still exists in our society and to the continued depression of women in our society, to the continued exploitation of poor and low-income people,*

---

568 Ian Oxenham, Prianka Ball, and Stephen O'Hanlon, "Op-Ed: Tri-College Divestment Campaigns; Growing Stronger Together," *Swarthmore Daily Gazette*, April 24, 2014. http://daily.swarthmore.edu/2014/04/24/op-ed-tri-college-divestment-campaigns-growing-stronger-together/

569 Nathan Livingston and Mark Parker Magyar, "Students and Trustees Debate Fossil Fuel Divestment," *The Spectator*, October 3, 2013. http://students.hamilton.edu/spectator/news/p/students-and-trustees-debate-fossil-fuel-divestment/view

570 *Fossil Fuel Divestment 101*, pg. 3.

NAS

and the continued exploitation in the global south by people in the global north. [571]

The solution was not "just cutting carbon emissions" to stop climate change but "fundamental restructuring of the political and social system for the people historically oppressed."

Some of that "restructuring" means rolling back historic exploitation. Many have begun to call for climate change reparations: mass payments of money, food, and foreign aid to compensate poor countries for suffering, or preparing to suffer, under the perceived ills of coming climate change. The first step, according to Tutu and McKibben, is to disassociate one's money from the industry that visits such climate change catastrophes upon the poor in the first place.

But climate change activism supposedly undertaken in the name of aiding impoverished Third World countries can harm the very people it intends to help. Developing nations need life-saving electricity and the benefits that come from industrial economies, not carbon offset payments to keep them from developing unused forests or stringing electric lines. Western carbon-phobia is already harming Third World nations by preventing the economic development that provides jobs, healthcare, and better food and water.[572] In the Kilwa region of Tanzania, for instance, peasants have invented a Swahili word, *njaa ya Bioshape*, for the bio-fuel starvation caused by the European craze to buy farmland and plant it with vegetation used to produce biofuels.[573] Those biofuels eat up crop land, drive up food prices, and make foreign food aid more expensive. Perhaps true climate justice requires giving up the anti-carbon hysteria.

Inside Divestment

Prima facie, the divestment movement aims to reduce American dependency on fossil fuels and to bankrupt the companies' political and social capital. But on a deeper level, divestment is only the smokescreen for heftier ideological ambitions. It is the *avant garde* of the sustainability movement, bringing in economic "equality" and social "reformation" along with its economic arguments against investing in fossil fuels.

William Lawrence, one of the Swarthmore Mountain Justice founders, views that ideological depth as a strong point for the divestment campaign. Unlike courses that teach students to spend time on old books and ideas rather than current urgent issues, and unlike programs that train students for careers that maintain the status quo's bent towards prejudice, Mountain Justice taught him to be an activist. He

---

571   William Lawrence, interview with Rachelle Peterson, November 12, 2014.

572   Caleb S. Rossiter, "Sacrificing Africa for Climate Change," *Wall Street Journal*, May 4, 2014. http://online.wsj.com/news/articles/SB10001424052702303380004579521791400395288

573   Chambi Chachage and Bernard Baha, "Accumulation by Land Dispossession and Labour Devaluation in Tanzania," International Land Coalition, May 2011. http://www.commercialpressuresonland.org/research-papers/accumulation-land-dispossession-and-labour-devaluation-tanzania

NAS

explained,

> *Lots of higher education institutions talk about some sort of commitment to the greater good.... To me, commitment to the greater good means campaigning for popular movements to transform the political, economic, and social structures of our country and world, to go from a system that works for the 1 percent to a system that works for the vast majority of people worldwide.*
>
> *I think an institution like Swarthmore is always a site of contest. One vision is that it is a comfortable institution that develops people for professional careers that maintain the status quo, and maybe do a good thing here or there. Another vision is an institution where people are able to get trained to become the sort of activists or organizers or healers that are needed to create the transformation. My ideal institution would be an academy for the incubation of social justice movements. That's not what Swarthmore is now. But I was fortunate to have something approximating that there, because there were a number of professors and students committed to that work. That's what my vision would be for highereducation.*[574]

Divestment treats the academy as a tool for political engagement. Not only must the school teach environmentalism, cut its own emissions, and encourage students to cut theirs. It must also conscript its very resources—endowments—into the service of sustainability.

In that regard, the divestment movement represents an especially egregious hijacking of the academy for political ends. It puts administrators—often themselves proponents of sustainability initiatives, and eager to cultivate environmental values in their students—in the uncomfortable position of attempting to uphold institutional financial stability while also remaining true to their own sustainability principles. It's a delicate balance that administrators are struggling to maintain—not because they misunderstand divestment, but because they understand all too well that their students are merely taking to the next logical extension the principles of sustainability that they themselves have nurtured. Sustainability has pushed hard into the heart of the college, and only now are those in the midst beginning to feel the pinch.

How long can such a balancing act go on, between financially-liable administrators and starry-eyed activist students? Not too much longer. As Bill McKibben said, do the math.

---

574   William Lawrence, interview with Rachelle Peterson, November 12, 2014.

NAS

CONCLUSION

The sustainability movement continues to develop. We have presented in these chapters an account of what it looks like as of the fall of 2014 and where it appears to be going. In this brief final chapter we venture some thoughts about its longer-term trajectory and offer our counsel on how higher education should respond.

The National Association of Scholars in recent years has published studies on a range of issues in higher education, including freshman courses in U.S. history at public universities in Texas (*Recasting History*, 2013), the disappearance of Western civilization survey courses (*The Vanishing West*, 2011), an annual study of common reading programs (*Beach Books*, 2010, 2011, 2013, 2014), and an in-depth study of one private liberal arts college (*What Does Bowdoin Teach?* 2013). One thing we have learned from the public reception of these studies is that many readers glance at the beginning and the ending of a report, and skip over the substantive middle. Readers are, understandably, in a hurry to know what the recommendations are. And some readers are all too willing simply to infer the content of a study from those recommendations.

We urge readers to forgo that approach. The value of this study resides in the detailed depiction of a social movement in the midst of self-creation. Our recommendations are of less consequence than our observations. But we do have some recommendations, in the form of advice to colleges and universities to uphold with greater vigor their traditional standards.

Respect Intellectual Freedom

Recommendation 1: Create neutral ground. Uphold the principle that higher education is neutral ground in important scientific debates. Colleges and universities betray something fundamental when they take sides in a dispute where there are serious differences on the key facts and pertinent theories. Some matters are indeed settled in history, science, or other fields—settled at least until an important new challenge arises.

But manmade global warming is not one of these. No college or university should have pre-empted this debate by signing the American College and University Presidents' Climate Commitment. Global warming caused by human agency may exist and may be important, but it is not the role of colleges and universities to declare the matter to be settled when it is plainly not. And it is all the more important that colleges and universities resist this rush to judgment when it appears that there is a majority on one side. Truth is not determined by majorities. It is determined by evidence and demonstration. The theory of man-made global warming has not achieved that standard.

NAS

Supporting the principle of neutral ground at this point requires that colleges and universities do more than say they are open for debate on these matters. They must act to ensure those debates actually occur on campus.

Recommendation 2: Cut the apocalyptic rhetoric. Presenting students with a steady diet of doomsday scenarios and insisting that the time for inquiry is over, and that it is instead the time for action, undercuts the basis for liberal education. Panicking students, demoralizing them, or imbuing them with a sense that the only morally acceptable course is blind obedience to a cause robs them of the opportunity to develop a mature understanding of the world.

Recommendation 3: Maintain civility. Some student sustainability protests have aimed at preventing opponents from speaking. The sustainability movement and its deep attraction to apocalyptic scenarios prompts a sizable fraction of students and even some faculty members to adopt a radicalized perspective. They come to believe that steps such as interfering with the free speech of others, taking over meetings, and attempting to prevail by force are legitimate tactics. No college or university can indulge these infringements against civility without sacrificing its basic claims to being a place set apart for the life of the mind and the disciplined pursuit of truth.

Recommendation 4: Stop "nudging." Leave students the space to make their own decisions. We have documented the efforts by many colleges and universities to manipulate students into adopting views and habits that are aligned with the sustainability movement. Psychological manipulation of students to get them to conform to an ideology is unworthy of higher education, which ought to aim at freeing students to examine matters with intellectual clarity. Stop the arm-twisting to make every course a sustainability course. Part of the American College and University Presidents' Climate Commitment is the idea that "sustainability" should be suffused throughout the curriculum. At many colleges this mandate is followed up with a requirement that faculty members, regardless of discipline, report yearly on their efforts to advance sustainability in their classes. The infringement on academic freedom is patent, but this approach also does a serious disservice to students who are robbed of a fair-minded approach to the subjects they choose to study.

Uphold Institutional Integrity

Recommendation 5: Withdraw from the ACUPCC. Colleges that have signed the American College and University Presidents' Climate Commitment should withdraw in favor of open-minded debate on the subject. The Climate Commitment is a dogmatic statement that compromises the institutions that sign it. At its worst, it becomes an invitation to suppress academic and intellectual freedom on campus for both students and faculty members. It is also an on-ramp to squandering large amounts of money in an effort to reduce institutional "carbon footprints." We have nothing against colleges and universities investing in efforts to curb their utility bills. But the pursuit of drastic reductions in institutional carbon footprints appears to impose net costs greatly in excess to any savings on carbon-based heating and electric generation.

Recommendation 6: Open the books and pull back the sustainability hires. Bring financial transparency to the campus pursuit of sustainability. We were unable to find a single college or university in the United States that offers a forthright and reasonably comprehensive account of what it currently spends on sustainability. Many costs are hidden inside other categories and not broken out. Institutions of higher education routinely boast that their sustainability "investments" save money in the long-term. But they make these claims behind an opaque wall when it comes to accounting for actual costs. It is time to change that practice by presenting a de-mystified account of what sustainability really costs. We have documented in this study the rapid growth of administrative and staff positions in sustainability in colleges and universities. These are cost-drivers in the immediate sense but they also represent the institutionalization of advocacy. The more such positions are added the more difficult it becomes for a college or university to uphold free inquiry on disputed matters.

Recommendation 7: Uphold environmental stewardship. Campuses need to recover the distinction between real environmental stewardship and a movement that uses the term as a springboard for a broader agenda. The blurring of this line has served the interests of those who like to appropriate the good will of students and the general public towards environmental goals, such as clean air and water, with political goals far removed from environmental concerns. Sustainability advances a hard-core anti-capitalist agenda and a commitment to the goals of a myriad of identity-based grievance groups. Sustainability advocates are up front about these goals on campus, but colleges and universities typically abet them by presenting sustainability to alumni and the broader public as simply an invigorated form of environmentalism.

Recommendation 8: Credential wisely. Curtail the aggrandizement of sustainability as a subject. There should be no such thing as a sustainability department or major. Sustainability is not a discipline or even a subject area. It is an ideology.

Be Even-Handed

Recommendation 9: Equalize treatment for advocates. Treat sustainability groups on campus under the same rubric as other advocacy groups. They should not enjoy privileged immunity from ordinary rules and special access to institutional resources. They should not receive favoritism or privilege, and they should be held to the same standards of openness and inclusion as every other group.

Recommendation 10: Examine motives. Boards of trustees should examine demands for divestment from fossil fuels skeptically and with full awareness of the ideological context. Their examination ought to be informed as well by an understanding of the debates over energy policy, including the once widely credited idea of "peak oil"; the advancement of new, relatively inexpensive ways to extract oil from deep layers of shale; the renewed debate about bringing nuclear energy back as a viable option; the recent world-wide plunge in the price of oil, and the financial difficulties faced by producers of "alternative" energy. There is no reason why boards of trustees should not give earnest consideration to the arguments of those who call for divestment. But if they decide to open themselves to this debate, trustees should pay attention to the full range of responsible views. It is the responsibility of higher education to create the space and conditions for both sides to make their best arguments and advance their best evidence.

We of course hope that colleges and universities will act favorably on all ten of our recommendations. But we realize that isn't likely. The leadership of much of American higher education is in the hands of people who are fully committed to sustainability. Even among college presidents who have reserved judgment, the prevailing sense is not to risk the wrath of the sustainability advocates. So the question remains, what is the longer-term trajectory of this movement? Will it succeed in embedding itself in higher education for generations to come, as perhaps the latest iteration of the Romantic Movement that commenced in the early days of the Industrial Revolution? Or will it flame out, like Occupy Wall Street?

Our view is that the sustainability advocates are working hard to institutionalize their movement. This may seem to contradict their apocalyptic narrative. If the world is coming to an end, why set in place long-term institutional structures? Why build energy plants meant to last half a century? Pretty clearly, the apocalyptic narrative is needed to create a sense of urgency but just as clearly it is indefinitely deferrable and not taken all that literally by many of the sustainatopians.

The longer term trajectory of the movement is to settle in as part of the permanent politicization of American higher education. Sustainability is a doctrine that justifies closing off the campus to inquiry and opinion that does not suit the views of those who favor a post-capitalist, post-national future. In due time, it will settle out as an "old" idea that must be replaced by something fresher. But left to its own course, sustainability by the time it is retired as an ideology will achieve a vast deforestation of our rich intellectual and academic environment.

Worse still, the sustainability movement is cultivating a susceptibility in today's students for the allures of command economies and undemocratic forms of political control. At its heart, sustainability is opposed to freedom. It offers students an imaginary world where important decisions about how to use resources will be made by properly credentialed experts, not by citizens making their own choices. The anti-consumerist impulse in this vision marches side-by-side with a wish for authoritarian control. Sustainability advocates are never too clear on exactly what regime they would like to install to bring about sustainatopia, but they are united in the belief that leaving people free to govern themselves can only create a tragedy of the commons. Sustainability means, "Do not trust your neighbor. He will despoil the Earth. Trust us. We will save it."

Striking the balance between trust and distrust is always the deep problem in politics. Higher education is one of the best places for people to wrestle with it. Whom should we trust more: Our neighbors who may make some short-sighted decisions? Or visionaries to whom it has been revealed how exactly we should live to ensure the future of the planet? The debate belongs in the classroom. The folly we face is not that the sustainatopians are arguing their views, but that they have increasingly monopolized the space. They come not to debate, but to rule. By that standard we can gauge what they will bring should they attain even greater power beyond the campus.

APPENDIX I: LIVING WITH CLIMATE ORTHODOXY ON CAMPUS: ONE PROFESSOR'S TAKE

Caleb S. Rossiter

What's it like being just about the only skeptic on a university campus? At first glance that seems like an absurd question. Academics are skeptics by inclination and by profession. That is what we do, and what we have done since the first of Plato's students strolled with him in the grove in Athens named after the hero Academus. The grove of Academus was a fine choice for Plato's inquiries into what is good, what is just, and what is true, since it had long been a site for the veneration of Athena, the goddess of wisdom.

Plato was a skeptic. He demanded that his students prove any of their assertions through logical argument and evidence. Indeed, he had inscribed over the gate of the grove, "All are welcome who love geometry." He didn't mean by this that you had to know how to circumscribe a triangle. He meant that you had to accept the rules of geometry's logical proofs: if you can't prove it, you don't believe it, but if somebody proves it to you, you must accept it.

Thanks to the intellectual sanctuary that universities have given to the concept of the academy for a thousand years, skepticism has given us all modern science and medicine, as well as practical politics and economics. Without the demand for proof and the refusal to genuflect to authority and conventional wisdom that skepticism breeds, we would still be living under slavery, monarchy, feudalism, and colonialism. Galileo and Copernicus would never have convinced us that the earth is not at the center of universe, Einstein that time slows down on a moving object (giving us $E=mc^2$ and nuclear power), and Bohr that under quantum mechanics we are all just waves of probability (giving us modern chemistry and nano-physics). Without those discoveries, all resisted by reigning paradigms when first claimed, we would lack nearly every industry and invention that has brightened and dramatically extended human life.

So how has "skeptic" come to be a dirty word on campuses? Because it has become short-hand for anyone who questions, let alone rejects, the orthodoxy of man-made climate catastrophe. Of all the beliefs that are brought to the academy, the belief that industrial energy emissions are a threat to human survival is the only one that escapes Plato's rules. A decade ago my graduate students in quantitative methods for international affairs wrote papers describing the peer-reviewed models and statistical analyses cited by the UN's Intergovernmental Panel on Climate Change in support of its claims of emission-driven climate catastrophe. As I delved into the models and the analyses so I could grade those papers I was surprised by how weak they were, and how misleading was the IPCC's portrayal of their certainty. Since then I have filled the unsought but inescapable role of climate skeptic at American University in Washington, DC.

NAS

Costa Rican Carbon Credits

Driving to campus recently to meet with a professor who is hosting an upcoming campus forum on "climate change," I was stuck behind one of our university's shuttle buses. Covering the back of the bus was an advertisement with significant consequences for my students, my paycheck, and my role as an educator: *"American University protects Costa Rican forests to offset the carbon emissions it creates with travel."*

What this means is that the university is writing checks to somebody in Costa Rica who promises not to cut down trees in a specific area. The justification for this policy comes from the theory that the trees that would otherwise be logged will continue to absorb carbon dioxide so that it doesn't mount to the atmosphere and contribute to a warming that leads to planetary disaster.

I wanted to pound my head on the steering wheel when I saw that ad. Our money will be passed on to Costa Rican landowners or the Costa Rican government. Small farmers who want to clear and work that land are out of luck. Logging cooperatives that want to harvest and replant trees are out of luck. Students who hope for lower tuition and professors who hope for more pay are out of luck. And the policy will have minimal influence on the amount of warming gases emitted to the atmosphere (which probably don't cause much change, let alone catastrophe in any event), and absolutely no influence on the university's vaunted and silly goal of achieving "carbon neutrality by 2020."

Oy vey. There is no such thing as carbon-neutrality in a booming, carbon-based economy. The bus that had the advertisement on it is made of materials mined, processed, and transported with carbon-based power, as is the road it runs on and the traffic lights it stops at. Students, professors, and staff need to eat healthy food, drink clean water, get medical care, be housed, travel, read, and take part in cultural events. All of these things have been developed and continue to be produced with carbon-based power. Like the wind farms and solar panels the university also subsidizes so it can add up putative carbon savings that "offset" the coal, gas, and oil that are converted to the electricity that powers America and its universities, the forest credits will not reduce emissions of carbon dioxide and methane. Emissions will continue to rise, as they always do, with demand and hence with income. All the offsets in the world will only slightly slow the rate of growth.

Our university instituted the offset policy because our president is one of the 685 who have signed the "American College and University Presidents' Climate Commitment." Here it is in its entirety:

> We, the undersigned presidents and chancellors of colleges and universities, are deeplyconcerned about the unprecedented scale and speed of global warming and its potential for large-scale, adverse health, social, economic and ecological effects. We recognize the scientific consensus that

NAS

global warming is real and is largely being caused by humans. We further recognize the need to reduce the global emission of greenhouse gases by 80% by mid-century at the latest, in order to avert the worst impacts of global warming and to reestablish the more stable climatic conditions that have made human progress over the last 10,000 years possible.

There is scarcely a phrase, scarcely a word, in that statement, starting with the hubris implied in its title, that we know enough about the workings of the climate system to contemplate controlling it, that can withstand the intellectual scrutiny that freshmen should be able to bring to the table after their first semester. I count at least ten different claims in the statement that are exceedingly difficult, if not impossible, to assess with meaningful certainty with the data that we have. And that doesn't count the claim, stentorian but essentially trivial, of a scientific consensus that human-driven global warming is "real." That it's real has never been in dispute. Industrial gases with odd numbers of molecules vibrate because of their lack of symmetry, and this creates warming when they happen to vibrate at the same frequencies as infra-red heat leaving the earth. The problem is that science has been unable to do more than guess at the "sensitivity" of temperature to those gases as the heat melds into the wild, complex maw of an actual, rather than theoretical, climate system.

For 685 academic leaders to sign on to a multiplicity of tenuous theories as proven fact sends a decidedly anti-academic message to their students. The authoritative tone of the statement intimidates less confident students from conducting the same sort of skeptical questioning we are teaching them to apply to all other claims. As Peter Wood of the National Association of Scholars wrote when properly correcting my notion that votes by boards of trustees to divest from energy companies at least force students to grapple with the issues, "Students shouldn't be pressured into endorsing (a board's beliefs). I favor 'grappling' with issues, but it is hard to grapple with anything when the institution dictates the answer."

Climate Shenanigans

Consider these excerpts from the report of American University's Climate Action Project Team of professors, administrators and activists, which was formed to implement the presidents' climate commitment:

AU faculty members strive to epitomize the scholar-teacher ideal by studying climate change side by side with students. In 2009, an envoy of six students accompanied a faculty member to Copenhagen to participate in the climate change treaty dialogue.[575]

Funny, nobody invited me. I think we can conclude that our participants were promoting one view, and one view only. The university shows its hand about that view by sponsoring a "dark night" each year on

---

[575] "American University: Carbon Neutral by 2020," American University, May 15, 2010. http://www.american.edu/finance/sustainability/upload/American_University_Climate_Action_Plan_-5-14-10.pdf

NAS

the campus, when all the campus lights are turned off as part of an international initiative to promote controls on industrial gases. The "dark night" organizers hope to get a satellite picture out of it that shows the developed world looking "like Africa" for a night. Sadly, they miss the irony that Africans want their continent to look like the developed world at night, because all the houses have electricity, rather than the one quarter that have it today.

> *Sustainability curriculum and research are being catalogued in order to identify areas of strength and opportunities for enhancing sustainability course offerings and connecting faculty sustainability interests with student interests and campus sustainability projects.*

Sustainability is a claim, a perspective, but not an academic discipline. Evaluating claims of sustainability, and the very concept and evidence for and against its validity—now that would be an academic exercise. The office of sustainability sent a "Go green" memo to all professors that included the suggestion that we raise grades for students who take part in demonstrations for legislation limiting emissions of fossil fuels. The memo offered us financial rewards and public recognition in a list of "green" professors if we reported our adoption of various "sustainable" practices, including this one. The policy was only dropped after I protested it in an article in the student newspaper.

> *The university is striving to demonstrate distinction in graduate and legal studies. The Washington College of Law offers one of the most robust programs of international and comparative environmental law in the country.*

I attended a conference at the law school, where a DC lawyer who was suing an energy company on behalf of an Alaskan Indian village described his tactics. I asked him about the role of the well-documented North Atlantic Oscillation in the recent rise in Alaskan temperatures, and whether he had to introduce any proof in court that the sea level around the village would have been different had nobody in the world ever driven an SUV. He laughed and said no, and that this was why the lawsuit was so promising: following the lead of the Supreme Court, all American courts now consider carbon dioxide a pollutant, although it has no negative effects on people, and simply stipulate the wildest claims of the United Nations' Intergovernmental Panel on Climate Change.

> *The university strives to engage the great ideas and issues of our time through research, centers, and institutes. The Center for Environmental Filmmaking hosts an on-campus Environmental Film Series and sponsors an Environmental Short contest.*

I attended two of the Center's events. The first was a screening of Al Gore's documentary about climate change, introduced by a single commentator who announced that all of Gore's claims were backed by a

NAS

consensus of the world's finest scientists. As I have written at length,[576] *An Inconvenient Truth* was actually filled with the sort of illogic and misinformation usually reserved for the closing arguments of a histrionic lawyer desperate to sway a jury. At the second event, a professor in the environmental film department screened his documentary about misleading claims in nature movies, where supposedly wild behavior and remarkable treks by animals are actually staged shots with animals that were domesticated, trained, and dragged about in front of the camera. He ended, though, by saying that such deception is justified to influence public policy: "I don't mind pretending that our staged penguins are in the wild, if it helps us save the penguins from losing their icy habitats to global warming." Whoa! Every last one of the millions of penguins in the world live in the Antarctic or very near it. In this region there has been very little warming, and actually a growth in ice in the past 50 years. Rest easy, friends of penguins.

Skepticism Affirmed

There has certainly been a positive side to being the one skeptical professor on campus, and becoming steeped in the physics, mathematics, and policies of the energy debate. My joint appointment in international studies and mathematics has afforded me wonderfully varied opportunities to engage students in thinking about how we come to believe things, and how we come to change our beliefs. In statistics courses, we use the IPCC's global temperature data to learn how changing the starting and ending periods of a time series changes our conclusions about trends. In math courses, we study the construction of climate models using differential equations, and observe how their projections for future years quickly "run away" exponentially to absurd conclusions, both boiling and freezing, unless they are "tuned" and arbitrarily curbed. In courses on African politics we study how developed countries' carbon-phobia blocks World Bank electricity projects, and promotes the seizure of farmland by European companies seeking biofuel credits.

Most important, the climate and energy debate provides real-world and real-time cases for students to use in pondering the questions of the skeptics in the original Academy. How do we come to believe something? How do we use logic to evaluate our beliefs? Do we have to defer to political and scientific authorities if we are not expert in their fields? What sort of evidence or proof does it take to change your mind, or someone else's?

I have seen environmental studies majors in shock, and at times in tears, in my class as they realize that there may be a fundamental flaw in their education. One told me, "I've been here three years and this is the first time I've even been exposed to the idea that there is uncertainty in the climate claims. I feel like I've been wasting my time. I have to rethink everything I've been believing." When I hear that, about any

576   Caleb S. Rossiter, "Climate Catastrophe: Convenient Fibs and Dangerous Prescriptions," CalebRossiter.com, March 2010. http://calebrossiter.com/Climate%202010.html

NAS

topic, I feel like I'm earning my paycheck. Rethinking, bringing skepticism to bear not just on others' claims but our own, is what we academics preach. It's nice to see it when it actually   happens.

So, getting back to the climate panel on which I'll be speaking soon, I'll be ever-optimistic. The faculty sponsor is a campus mentor to the students who have organized to demand divestment of the university's portfolio from energy stocks. We've been colleagues in social activism since 1984, when he was running a group that lobbied Congress to end U.S. support for dictators and civil wars in Latin America and I was a congressional staffer trying to do the same thing. This is the first issue on which we've fundamentally disagreed. When we met that day I saw the campus bus bragging about Costa Rican offsets he stared at me in horror and disbelief as I explained my conclusions from studying and teaching the statistics behind claims of climate catastrophe. But he's an educator, the real deal. He signed me up to broaden the discussion, saying that it would do the students, and him, good to think about the other side. As long as we have professors like him who are trying to educate rather than proselytize, we skeptics in the academy will do just fine.

Dr. Caleb S. Rossiter is an adjunct professional lecturer at American University in the School of International Service and the Department of Mathematics and   Statistics.

APPENDIX II: DIVESTING REASON: A STUDENT'S PERSPECTIVE

Danielle Charette

As a student at Swarthmore College, where activists like to claim the national fossil fuel divestment campaign got its start, I observed two contradictory phenomena. The divestment movement is, at once, utterly nonsensical and totally appealing to collegiate do-gooders.

NAS in this report persuasively demonstrates why divestment defies common sense. Classmates of mine liked to argue that pulling Swarthmore's $1.5 billion endowment out of the fuel industry would set off a "domino effect." That is, other universities and non-profits would be inspired or peer-pressured into rethinking their own investments. In time, the thinking goes, oil and gas companies will lose both their profitability and credibility. Of course, even if Swarthmore were to renounce its shareholder influence and sway other institutions to do the same, the notion that divestment would spur a "domino effect" is utopian thinking. For starters, it assumes that most Americans look to elite liberal arts schools for their financial advice. They don't. While some friends and relatives were impressed that I had beaten Swarthmore's admissions odds, an equal number scoffed at the College's $59,610 (and rising) annual price tag. Between the explosion of overpaid administrators, needless development projects, and the persistence of antiquated tenure policies, private colleges like Swarthmore elude the business-minded approach. Aspiring CEOs will probably not be consulting the Swarthmore Board of Managers for financial tips.

But this brings me to my second point: When it comes to appealing to the typical campus activist, divestment is brilliant. Any cogent objection to divestment is likely to be financial. But as soon as the conversation turns to stock prices and dividends, it's easy for campus organizers—many of whom are already quite suspicious of capitalism—to portray their opponents as greedy number-crunchers. Despite the College's exorbitant tuition, the *Princeton Review* consistently ranks Swarthmore one of its "best value" colleges thanks to generous financial aid. So our conservative club, which I co-led, asked classmates what they thought divestment's effect on aid would be, especially since aid is by far the fastest growing item in the College's budget. Our peers responded that we were presenting a false choice. We were invoking the economic maxim of scarcity; they were talking about social justice.

This perceived disconnect between economics and social justice is why student protestors felt vindicated in taking over a 2013 Board of Managers meeting. The May meeting was originally coordinated between the administration, board, and representatives from Swarthmore's pro-divestment group, Mountain Justice. Our president advertised the meeting as a civil forum for students, faculty, and administrators on all sides of the divestment issue to express their views. But rather than listen to chairman of the Board

NAS

Investment Committee give his opening PowerPoint presentation, Mountain Justice activists snatched the microphone out of turn. As I detailed in a *Wall Street Journal* op-ed,[577] the chairman's presentation was usurped by a series of protestors who condemned the "liberal script" in the name of "radical, emancipatory change" and "institutional transformation."

I was in the audience to support a friend of mine, who was hoping to offer a few remarks critical of divestment during the moderated discussion period. But now that the moderator had left her post at the front of the auditorium and ceded the podium to a long line of angry activists, we realized no such level-headed discussion would be taking place. I was so startled that neither the College president nor the dean of students nor any of the other administrators in the audience made any effort to regain order that I stood up and asked for a return to the advertised format. Yet every administrator remained passive and shell-shocked, as protestors began to clap in unison, drowning out my pleas for order. In a bizarre inversion of authority, Swarthmore's then-president Rebecca Chopp agreed with me that what was unfolding was "outrageous," yet she said there was nothing she could do. Afterwards President Chopp informed me that the administration has a policy of resorting to Quaker-styled silence when aggressive protests break out, to demonstrate that they are listening to student concerns. But remaining mum in the face of student unrest assumes that all students hold the same views. In a university environment where students are bound to disagree—sometimes aggressively so—such passive leadership only fuels mob rule.

Nevertheless, Swarthmore professors like George Lakey, who teaches in the Peace and Conflict Studies program, praised the takeover as an "egalitarian" invasion "of the 1 percent's space." Furthermore, Lakey wrote that the "conflict-aversion" displayed by those of us who believe in "civil discourse" is no more than "an ally of the 1 percent because it keeps people apart and solidifies the status quo."[578]

Clearly Professor Lakey is talking about more than the issue of divestment here. Anyone at that Board Meeting would have heard protestors raise a number of seemingly unrelated grievances, from sexual assault to immigration issues to alleged hate crimes. The environmentalists who originally organized the meeting were more than happy to have divestment serve as a mere introduction into an entire worldview where the "marginalized" speak truth to power. Students at liberal arts colleges like Swarthmore hear a lot about "intersectionality"—or the relationship between all systemic forms of "oppression." Intersectionality is a theory that reduces society to a battle between "us" and "them." On one side are activists who crusade for "justice" and on the other are all who stand in their way.

---

577   Danielle Charette, "My Top-Notch Illiberal Arts Education."

578   George Lakey, "Swarthmore College's Rude Awakening to Oppression in its Midst," *Waging Non-Violence*, May 21, 2013. http://wagingnonviolence.org/feature/swarthmore-colleges-rude-awakening-to-oppression-in-its-midst/

NAS

Herein lies divestment's fundamental appeal: It purports to address serious global concerns by means of a local target: college administrators and fellow students. Suddenly, academic bureaucrats who refuse to divest their endowments are the face of worldwide injustice. Protestors find that they can participate in the thrills of civil disobedience—whether that means overthrowing school meetings or leaving fistfuls of coal in board members' mailboxes. And as the Swarthmore administration's hapless response to the meeting takeover demonstrates, campus disobedience is relatively safe. It is free from the kind of disciplinary risks one might face for, say, storming the Senate floor. Activists don't need to persuade the American public that divestment makes sense—just the classmates who have been schooled in the same rhetoric of intersectionality and social justice.

By keeping things local, divestment activists catch their own administrators flat-footed. College presidents wax poetic about "sustainability" and brag about how many academic courses are cross-listed with the environmental studies program, yet they draw the line at divestment. Divestment proponents point out— usually quite loudly—that this is hypocritical. Shocked to find themselves framed as conservative oppressors, college administrators—most of whom pride themselves as good progressives—are too ambivalent to offer divestment the vigorous rebuke it deserves.

That rebuke should include economic arguments, but educators will never persuade young leftists by simply appealing to the laws of supply and demand. Equally important is a defense of the free exchange of ideas. When demonstrators resort to grabbing microphones and "clapping down" their peers, what they really demonstrate is the shallowness of their arguments. One of the reasons endowments like Swarthmore's are so large is that successful alums believe the liberal arts are worth defending. We have a responsibility to honor their investment.

Danielle Charette graduated from Swarthmore College in 2014 with a BA in English Literature. She was the co-founder of the Swarthmore Conservative Society. She is now a Ph.D candidate in the University of Chicago's Committee on Social Thought.

APPENDIX III: ENVIRONMENTAL LYSENKOISM: REFLECTIONS FROM A SCIENTIST

William Happer

Andrei Sakharov, the father of the Russian hydrogen bomb, later justly revered for his work on human rights, describes the following incident at "the object," Russia's nuclear weapons laboratory Arzamas-16:

> In the middle of the year 1950 a committee arrived at the object, perhaps from the Principal Administration, or maybe from somewhere else, to check out the leading scientific cadres. We were called one by one to the committee. They asked me several questions, which I can't remember; afterwards there was the following question:
>
> 'What is your assessment of the chromosome theory of inheritance?'
>
> (This was after the 1948 session of the Lenin All-Union Academy of Agricultural Sciences, when Stalin sanctioned Lysenko's destruction of genetics. Consequently, the question was a loyalty test.) I answered that I considered the chromosome theory scientifically correct. The members of the commission exchanged glances but said nothing. No organizational changes of my position followed. Evidently, my position and role at the object were already sufficiently strong, that it was possible to ignore sins of this nature.[579]

Sakharov's endorsement of modern genetics directly contradicted an alternative Soviet theory, Lysenkoism, which held that acquired traits could be passed on from generation to generation. Lysenkoism had no scientific standing, but Joseph Stalin officially endorsed it. In the Soviet Union, affirming Lysenkoism had become, as Sakharov put it, a "loyalty test."

A few weeks after Sakharov was questioned, a colleague, Altshuler, another distinguished nuclear physicist, came under questioning as well:

> A couple weeks later, Zeldovich came to me and said that we had to help out Altshuler.... It turns out that Altshuler was given the same question that the committee had given me, and with his characteristic straightforwardness, he answered just as I had, but unlike me, he was threatened with dismissal.[580]

After Sakharov pleaded for Altshuler, the secret police deputy at Arzamas-16 told him:

---

579   Andrei Sakharov, *Vospominaia* (Memoirs), New York: Chekhov Publishing House, 1990, pg. 181. Translation by William Happer.

Yes, I already heard about the hooligan-like outburst of Altshuler. You say that he has done a lot for the object and will be useful for further work. Right now we won't make any organizational changes; we will see how he behaves in the   future.[581]

Nine years earlier, in 1941, Nikolai Vavilov, a world-renowned biologist, had been sentenced to death for maintaining that chromosomes had something to do with inheritance. But by 1950, the Soviet leadership, most notably Stalin and Beria, was so desperate to break the USA's monopoly on nuclear weapons that they reluctantly forgave Sakharov's heresy. After all, he had invented the Soviet hydrogen bomb, a very successful design, better in some ways than the rival American design. Who knew what he might invent next? But scientists without this unusual protection continued to be at high risk of losing their jobs, or worse, if they expressed any support for the imperialist myth of genes and chromosomes.

Lysenko was a poorly-educated agricultural extension agent from Ukraine, but in the turbulent early years of the Soviet Union he managed to convince the political leadership that Mendelian genetics and hybrid vigor were an evil imperialistic fiction. He maintained that living organisms could inherit acquired characteristics, and that with the right proletarian science the Soviet Union could enormously increase agricultural yields. Lysenko violently opposed the introduction of hybrid corn to the Soviet Union, ostensibly because it was a fairy tale invented to enrich capitalist seed companies. (Today we hear much the same propaganda about genetically modified crops from Greenpeace, and from many European governments.) Despite the fact that none of his theories were ever validated by experiments, Lysenko flourished in the Soviet Union, starting in the early 1930s under Stalin, until he shared in Krushchev's downfall in 1964. Lysenko orchestrated a 40-year reign of terror from which Russian study of biology has not fully recovered to this day.

Lysenko laid out his theory in a report canonized during the infamous 1948 session of the Lenin All-Union Academy of Agricultural Science. His final words there indicate the deep link between Soviet science and politics:

> *In one of the notes I am asked, what is the attitude of the Central Committee of the party to my report? I answer: the Central Committee of the Party has looked over my report and approved it.*[582]

After this the stenographer's notes record: "Stormy applause growing to a standing ovation."[583]

Reminiscing about the 1948 report that destroyed the careers of so many honest scientists, Lysenko

---

581   *Ibid.*

582   Valerie Soifer, *Vlast' I Nauka* (Power and Science), Tenafly NJ: Hermitage, 1989, pg. 410. Translation by William Happer.

583   *Ibid.*

NAS

wrote a short obituary note in the Russian Communist newspaper *Pravda* following Stalin's death:

> *Stalin, the Guiding Light of Science: Stalin ... directly edited a draft of the report, About the Situation in Biological Science. He explained in detail his corrections, gave me hints on how to present various parts of the report. Comrade Stalin attentively followed the results of the work of the session...*[584]

Many Soviet citizens, philosophers, journalists, and those eager to be admitted to the Communist party joined the Lysenko bandwagon.

Surely something like Lysenkoism cannot happen in the United States or the West? Alas, it has already happened. Fanatical environmentalism has adopted many of the methods of Lysenkoism. Being sentenced to death for scientific heresy has been out of fashion in the West for a few centuries. But that has not stopped calls to execute "deniers" who question the dogma that CO2 from the combustion of fossil fuels will cause catastrophic global warming. The very word "denier" is cynically used to evoke the image of a Holocaust denier, a neo-Nazi. During a 2014 campaign speech in California, President Obama sneered, "So unfortunately, inside of Washington we've still got some climate deniers who shout loud, but they're wasting everybody's time on a settled debate." Just as Stalin considered himself an expert on biology, there are lots of people, including President Obama, who consider themselves experts on climate science, despite their lack of formal training.

Woe to that American scientist today who suggests that neither basic theory nor observations support the politically-correct dogma that more CO2 in the atmosphere will bring on the four horsemen of the apocalypse. Pestilential tropical diseases will spread toward the poles; nations will go to War over the remaining resources of a world blighted by the demon gas, CO2. Famine and Death will pick up the pieces. Even mild disagreement with these absurd claims can ruin an academic   career.

Especially alarming has been the complicity of academia with environmental extremists. There are many egregious examples. A case in point is the treatment of two climate scientists at the University of Virginia at Charlottesville: Dr. Patrick Michaels, a global warming skeptic, and Dr. Michael Mann, who created the celebrated, and much-disputed, "hockey-stick" temperature graph for the past 1000 years.

After Dr. Michaels left the University of Virginia to join the conservative Cato Institute, Greenpeace demanded copies of Dr. Michael's e-mails. The university fell over itself to comply with Greenpeace. University policies stipulate that e-mails are to be destroyed 30 days after an employee's last pay check. But not only were Dr. Michaels' files not destroyed, as required, but the university was eager to use the

---

584   *Ibid.*

illegally retained files to help Greenpeace smear its former employee.

After Michael Mann left the University of Virginia to accept a professorship at Pennsylvania State University, the University of Virginia received an analogous freedom of information request from a Virginia state legislator to inspect Dr. Mann's e-mails. The university responded that since Dr. Mann was no longer an employee, the e-mails had been destroyed. The university squealed indignantly that they and their politically-correct former employer were the victims of a witch hunt.

Dr. David Legates had a very similar experience at the University of Delaware, where the university administration went out of its way to honor a Greenpeace demand to see all of his e-mails related to global climate change and similar topics. Requests from the Competitive Enterprise Institute to see e-mails of another University of Delaware professor, whose views supported global warming alarmism, were curtly dismissed by the university's General Counsel. Disturbing details of this incident can be found in Professor Legates' testimony to the Public Works Committee of the United States Senate on 3 June, 2014.[585] Dr. Willie Soon of the Harvard Smithsonian Observatory and Dr. George Taylor of Oregon State University are just a few of the many other honest and innovative scholars who have suffered for opposing the demonization of carbon dioxide, just as Lysenko's victims suffered for opposing the demonization of Mendelian genetics. While Michaels, Legates and many others are persecuted, their politically-correct colleagues are showered with research grants, academic prizes, elections to scientific academies, and other rewards. Much the same thing happened in the Lysenko area of the Soviet Union. The irony is that more CO2 is probably going to be good for the planet. Observations show clearly that the warming potential of more CO2 has been grossly exaggerated. The modest warming from doubling or tripling current CO2 concentrations, along with the enormous benefits to agriculture from this life-giving, essential molecule, will be a benefit to the planet and to humanity.

Most people are instinctive, sensible environmentalists. They want to protect the beautiful world we live in. But they want to be permitted to live decent lives themselves. But the only way for humans to have no impact on the environment is for them to disappear from planet Earth. In its influential 1991 report, *The First Global Revolution*, the Club of Rome wrote,

> *In searching for a common enemy against whom we can unite, we came up with the idea that pollution, the threat of global warming, water shortages, famine and the like would fit the bill. In their totality and their interactions these phenomena do constitute a common threat which must be confronted by everyone together. But in designating these dangers as the enemy, we fall into*

585 David R. Legates, "Statement to the Environment and Public Works Committee of the United States Senate," United States Senate, Environment and Public Works Committee, June 3, 2014. http://www.epw.senate.gov/public/index.cfm?FuseAction=Files.View&FileStore_id=aa8f25be-f093-47b1-bb26-1eb4c4a23de2

NAS

the trap which we have already warned readers about, namely mistaking symptoms for causes. All these dangers are caused by human intervention in natural processes, and it is only through changed attitudes and behavior that they can be overcome. The real enemy is then humanity itself.[586]

But the real danger to environmental scholarship is the dearth of scientists like Michaels, Legates and others who follow the evidence rather than the politics, and the overwhelming pressure for researchers to conform to an ideology. Environmental Lysenkoism not only threatens the soundness of public policy. It assaults the trustworthiness of science itself.

William Happer is the Cyrus Fogg Brackett Professor of Physics at Princeton University. From 1991-1993 he was the Director of the Office of Energy Research at the U.S. Department of Energy.

---

586   Alexander King and Bertrand Schneider, *The First Global Revolution*, Council of the Club of Rome, Orient Longman, 1991, pg. 75.

APPENDIX IV: A TRANSNATIONAL, "PRECAUTIONARY" MOVEMENT: THOUGHTS FROM AN INTERNATIONAL TRADE LAWYER

Lawrence A. Kogan, J.D.

The global sustainability movement has gradually imposed its tenets and strictures upon, and steadily asserted greater control over the economic lives of, national citizenries directly and indirectly via enactment of governmental 'hard law' and social 'soft law' norms. This cross-border movement has successfully developed, prompted or otherwise triggered:

1. national governmental adoption of international treaties and public standards;

2. national and state governmental and environmental group legal actions commenced at international and/or national tribunals;

3. governmental promulgation of federal, state and municipal laws and ordinances, especially those public procurement-related;

4. industry adoption of corporate social responsibility and sustainability mandates and standards through reputation-harming public 'naming and shaming' campaigns and campaigns of physical intimidation; and

5. public shareholder and boardroom activism at the hands of environmental activist groups and state and local government pension and investment funds.

In addition, the global sustainability movement has employed non-legal means, including manipulation of language, media campaigns and moral suasion to forge a new social compact rooted in political consensus. The National Association of Scholars, in this well written and documented report, has discussed how the campus sustainability movement has emulated many of these strategies and tactics.

At the fulcrum of the sustainability movement is the "precautionary principle," a well-recognized ostensibly common sense-based "better-safe-than-sorry" legal nostrum incorporated in European constitutional and civil law treaties and regional environmental, health and safety laws. European governments deem the preemptive features of the precautionary principle necessary to address what they consider the "unknown-unknowns." The precautionary principle also is, perhaps, among the most subtle legal concepts associated with the global sustainability movement because it masks policy-based science as science-based policy, and enables governmental authorities to evade calls for regulatory transparency and accountability.

Regulation and Evidentiary Thresholds

Governments, in other words, have employed the precautionary principle politically in the name of science as a preemptive palliative to eliminate the perception of risks posed by the everyday use of

NAS

substances and products and the engaging in everyday activities. It has been effectively invoked *a priori* in the absence of quantifiable empirical causal lines of scientific evidence of observed or observable risks of harm posed by actual or historical use, dosage, and exposure to refocus attention on the intrinsic qualitative characteristics of a substance, product, or activity without regard to use, dosage, or exposure. To this end, global sustainability advocates employ the precautionary principle to emphasize, for regulatory purposes, mostly subjectively weighted correlative evidence of unknowable possible or potential future environmental health and safety hazards, rather than probable "known-known" or "known-unknown" environmental, health, and safety risks that substance and product uses and activities undertaken actually engender. And, they do so largely without regard to the economic costs and burdens imposed on domestic and foreign economic actors.

To better accommodate governments' desire for more frequent invocation of the precautionary principle, progressive European and American scientists have successfully reconstituted the international metrics of risk assessment so that they are expressed in qualitative as well as quantitative terms. As a result, application of the precautionary principle results not only in the lowering of scientific evidentiary thresholds for identifying the existence of health and environmental harm from causation to correlation, but also in the reduction of the legal evidentiary standards and a shifting of the burden of proof (from the government to economic actors) necessary for triggering ex ante as well as post hoc governmental regulatory actions.

Exporting Precaution

The European Union has long endeavored to export the precautionary principle to many of its international trading partners for purposes of establishing it as an absolute international legal norm from which no derogation of adherence would be tolerated, even among non-parties to international treaties that incorporate it. Industries within many nations, however, have criticized this effort as an extraterritorial imposition of cultural preferences or a form of cultural imperialism, and as disguised regulatory trade protectionism designed to level the economic playing field in favor of domestic European industries besieged with the costs and burdens of unilaterally imposed precautionary principle-based regional regulations. Although the United States joined this effort only recently because of the Obama administration's predisposition toward importing from Europe and developing its own precautionary principle-based regulations, it has since been similarly and justifiably criticized.

Media reporting surrounding recent United Nations climate change negotiations indicates that the EU and the U.S. have encountered stiff political resistance from emerging economies such as India and China. These countries have rejected as unreasonable the called upon forbearance of economic development "as we know it" that the execution and implementation of a proposed new post-Kyoto climate treaty premised on the precautionary principle would require. Perhaps, if these countries were reassured that the scientific assessments supporting the UN Intergovernmental Panel on Climate Change ("IPCC")'s

Fourth and Fifth Assessment Reports upon which current climate change treaty negotiations are premised had been properly peer reviewed and scientifically validated pursuant to the provisions of an enforceable government transparency and accountability mechanism, such as the U.S. Information Quality Act ("IQA"), they would be more inclined to seriously consider such a treaty. However, such a result would depend on the U.S. government, itself, conforming to IQA statutory and administrative standards in connection with federal agency use of nationally-developed climate assessments as support for environmental regulations – which the U.S. government is neither willing nor able to do.

ITSSD

The Institute for Trade, Standards and Sustainable Development (ITSSD) is a nonprofit legal research and analytics organization that promotes a positive paradigm of sustainable development that affords future generations from all sovereign nations greater opportunities for a higher quality of life. To achieve this positive paradigm we emphasize the importance of free markets, free trade, economic growth, the rule of law, strong tangible and intangible private property rights, scientific discovery, and technological innovation. We also emphasize the need to ensure governments' open and transparent establishment, maintenance, and oversight of balanced, risk-based science, and economic cost-benefit analysis-driven national regulatory and standards schemes, and the quality and integrity of scientific & technical data/ information that government entities rely upon, adopt as their own and disseminate to the public as a basis for agency actions, including rulemakings.

During the past decade, ITSSD has endeavored in the public interest to identify, examine, and report the emergence of a global sustainability movement that has developed considerably since the choreographed World Trade Organization—Seattle protests of 1999, when pro-labor, anti-globalist protesters blockaded intersections and prevented international delegates from arriving at the conference. This sustainability movement has been advanced by foreign and domestic environmental groups, European national and regional governments, and more recently, by U.S. federal, state, and local government officials. The sustainability movement's objective is to forge a radically new global social, political, economic, legal and ethical order and consciousness–a paradigm-shift of transformational proportions–focused on the alleged collective need and urgency to modify present individual human behaviors for the putative benefit of future generations.

At its core, this new paradigm is 'post-modern.' This means it is fundamentally antithetical to Enlightenment-era humanism and its societal, scientific, economic, legal and political institutions and ideals. Sustainable development is a progressive, "social democratic" framework connected to European social norms that are, in their most extreme form, anti-anthropogenic. They veer towards and sometimes cross the line between advocating reform of modern society and calling for its uprooting and destruction.

NAS

Indeed, ITSSD's work has focused on identifying and assessing the systemic risks that the concept of "negative" sustainable development and a borderless global sustainability movement pose to national sovereignty, common law notions of private property ownership, individual economic and political freedom, and the rule of law. These are indispensable Enlightenment era natural rights-based principles that undergird this nation's founding which are embedded in the U.S. Federalist Constitution, Bill of Rights, and Declaration of Independence.

ITSSD, the EPA and NOAA

ITSSD has recently focused its research and reporting efforts to identify and highlight the extent of U.S. federal agency compliance with the Information Quality Act in connection with the publicly disseminated scientific assessments underlying new and recently proposed environmental regulations implementing inter alia the U.S. Clean Air Act. In May 2013, ITSSD filed an *amicus curiae* brief in the United States Supreme Court in the case of Coalition for Responsible Regulation vs. EPA which requested judicial review on such grounds. If ITSSD's request had been granted, the Court would likely have required the reexamination of EPA's prior peer reviews of the third-party climate assessments supporting its 2009 Clean Air Act GHG Endangerment Findings.

Since the Court did not grant judicial review of the case on this issue (because it had not been adequately raised by the litigants in the lower court), ITSSD filed, in March and April 2014, detailed and annotated Freedom of Information Act ("FOIA") requests with the U.S. Environmental Protection Agency ("EPA") and the National Oceanic and Atmospheric Administration ("NOAA"). In particular, these FOIA requests sought public disclosure of agency records substantiating that the many third-party-developed climate assessments EPA had adopted, used and publicly disseminated as the scientific foundation for the EPA Administrator's 2009 Clean Air Act Section 202(a) Greenhouse Gas Endangerment Findings, including numerous NOAA-developed climate assessments, had been peer reviewed in conformance with the IQA's most rigorous and least discretionary peer review, transparency, objectivity, independence, and conflicts-of-interest standards applicable to "highly influential scientific assessments" ("HISAs").

To date, neither agency has substantively responded to these FOIA requests, other than to request, in return, that ITSSD redraft them more narrowly. In June 2014, ITSSD filed with EPA a more detailed and annotated FOIA request. In July, it received in response a request for payment assurance in the amount of $27,000 evidencing EPA's prospective assessment of search fees for processing ITSSD's FOIA request "as-is", along with a rejection of ITSSD's request for a statutory fee waiver. In August, ITSSD filed its Appeal of that rejection with the EPA's Office of General Counsel. Those interested in following ITSSD's IQA-focused FOIA activities and the institutional and media reporting and editorials discussing them may access the ITSSD website at: www.itssd.org.

NAS

Support for NAS

ITSSD clearly has been developing a critique of the sustainability movement that differs in important ways from the National Association of Scholars' critique. ITSSD focuses on the governmental and regulatory side of things; NAS on higher education. Independently of one another, however, ITSSD and NAS have come to many of the same conclusions about the illiberal and ideological character of this movement, and its determination to by-pass standards of transparency and public accountability. I am very pleased to partner with NAS in the effort to document and analyze the sustainability movement and to bring it to a higher level of public scrutiny.

Lawrence A. Kogan, J.D., LLM is Managing Principal of Kogan Law Group, P.C., and CEO of the nonprofit Institute for Trade, Standards and Sustainable Development

NAS

INDEX

NAS

NAS

NAS

NAS

NAS

8 W 38TH ST. SUITE 503   NEW YORK, NY 10018   917.551.6770   www.NAS.ORG

www.ingramcontent.com/pod-product-compliance
Lightning Source LLC
Chambersburg PA
CBHW061359210326
41598CB00035B/6039